JLC's Construction
Tips & Techniques

Practical Answers to Contractors' Questions

From the Editors of

The Journal of Light Construction

With contributions from Michael Byrne,
Rex Cauldwell, Henri deMarne, Paul Fisette,
Henry Spies, and others

A Journal of Light Construction Book

www.jlconline.com

hanley▲wood

Editors: Clayton DeKorne, Don Jackson, Sal Alfano
Managing Editor: Josie Masterson-Glen
Proofreader: Emily Stetson

Graphic Designer: Terry Fallon
Illustrators: Tim Healey, Chuck Lockhart
Cover Design: Kimberly O'Connell
Production Director: Theresa Emerson
Cover Photo: Bryan Striegler

International Standard Book Number
ISBN-13: 978-1-928580-31-7
ISBN-10: 1-928580-31-9
Library of Congress Catalog Card Number 2006930870
Printed in the United States of America

A Journal of Light Construction Book

The Journal of Light Construction is a tradename of Hanley Wood LLC.

The Journal of Light Construction
186 Allen Brook Lane
Williston, VT 05495

Table of Contents

Introduction

On the typical residential job site, hardly a day goes by without some question being raised. How could we do that faster? Is there a better way? What's the strongest — or most durable — method? How do I install this product?

Even after a project is built, the questions continue. Why is there condensation in the attic? What's causing those black spots? Why is the exterior finish flaking?

This book gathers together questions that builders have sent JLC over the past decade, and provides answers written by the most capable experts we could find. We weren't able to include all the questions we've received, so we chose ones that seemed most representative of the common experience of builders everywhere.

Many of the questions have simple, straightforward answers. But some of the answers are less clear-cut — attempts to troubleshoot perplexing situations, to arrive at a plausible explanation for a baffling set of conditions. For me, these are among the most interesting: they don't so much tell you *what* to think as *how* to think about unexpected results on the job.

We hope you enjoy the book and find it useful. And as new questions arise, we hope you'll ask us. We welcome every opportunity to do what we do best: provide sound, practical information you can build by.

Don Jackson
JLC Editor

Foundations & Sitework

1

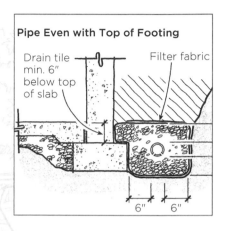

Pipe Even with Top of Footing

Drain tile min. 6" below top of slab

Filter fabric

6"   6"

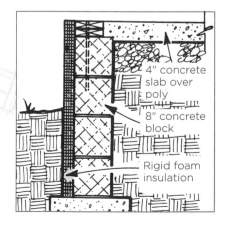

4" concrete slab over poly

8" concrete block

Rigid foam insulation

Orienting a Foundation Drain

Q. Should the holes in a foundation perimeter drain face up or face down? And does the drain need to be pitched as it runs around the house?

A. There's no question that the holes must face down (oriented approximately at 4 o'clock and 8 o'clock) to drain off any water rising from below. Keep in mind that if the drain is embedded in stone, as it should be, water will run through the stone as well as through the drain. In other words, the stone and pipe work together to drain the water away. You aren't limited to just the narrow trough of water between the holes in the pipe.

The drain should be placed so the top of the pipe is at least 6 inches below the interior slab. If the holes were facing up, you might be able to carry off more water through the pipe itself, but in order to get into the pipe, the water level would rise dangerously close to the slab.

Ideally, the drain should pitch as it runs around the house, but often this isn't practical. To get a $1/8$-inch-per-foot pitch, one corner of the drain on a 20x30-foot house would have to be 6 inches above the opposite corner, bringing the drain above the slab. To avoid this, the

Pipe Even with Top of Footing

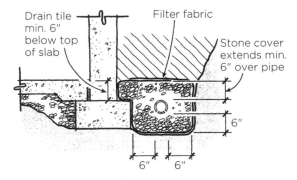

Drain tile min. 6" below top of slab — Filter fabric — Stone cover extends min. 6" over pipe — 6" — 6" — 6"

Pipe at Bottom of Footing

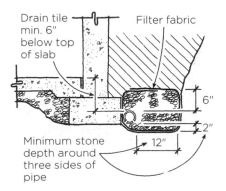

Drain tile min. 6" below top of slab — Filter fabric — 6" — 2" — Minimum stone depth around three sides of pipe — 12"

The best location for rigid drain pipe is alongside the footing. Ideally, the drain should be at least 6 inches below the top of the slab and always covered by at least 6 inches of stone.

drain can be laid level around the house, but it must not have any dips or rises, and it must pitch away from the house to daylight. A pitch of $1/8$ to $1/4$ inch per foot is recommended. — *Don Marsh*

Waterproofing vs. Damp-proofing

Q. There seem to be a lot of products on the market that solve problems with leaky basements. Some are for "damp-proofing" and others for "waterproofing." What is the difference?

A. A product that fills the pores of a material to reduce capillary movement of water through the material can only be called damp-proofing. An example is the thin bituminous coating that is painted on basement walls to reduce vapor flow through the wall. These products are useless when bulk water or saturated soil presses against the wall and exerts hydrostatic pressure.

By comparison, if the product produces a sealed membrane or forms a continuous film with essentially zero permeability (very difficult), it could be classified as waterproofing. The material must also be able to bridge cracks and be flexible enough to withstand movement. Examples include synthetic rubber membranes and bentonite-based products.
— *Henry Spies*

Sealing a Foundation Cold Joint

Q. We are pouring a foundation for a full-basement addition in an area with a high seasonal water table. What is the best way to detail the joint between an existing foundation and the new foundation to prevent water from leaking in?

A. Any sealant is just a band-aid, not a lasting solution to the problem. You have to deal with draining the water first. Then you can worry about sealing the joint between the old and new foundations with backer rod and urethane caulk to keep out radon gas, if necessary.

Here in northeastern Pennsylvania we have high seasonal water tables, so for all foundations we pour, we use an interior perimeter drain to deal with the water. Most builders think in terms of an exterior drain. But if you have a high water table, the water will rise inside the foundation. Unless you have an interior drain, the footings essentially act like a dam, preventing the water from reaching the exterior drain and forcing it

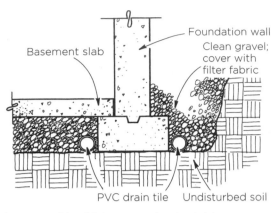

In areas with a high seasonal water table, a foundation should have both interior and exterior perimeter drains. Run the drains to daylight or to a sump pump connected to a storm drain.

to rise through the slab.

We typically form and pour our footings and lay drain line around the inside and outside perimeters (see illustration). The drains must run to daylight or to a sump pump connected to a storm drain.

Apply a standard asphalt-based foundation coating to the exterior of the foundation, then backfill with plenty of gravel, with filter fabric protecting the footing drains. Water is predictable and will always take the path of least resistance. This means if you backfill the outside of the foundation with enough gravel and provide a daylight drain, the water shouldn't rise along the foundation wall and won't have enough pressure to leak in through the joint.

— *Carl Hagstrom*

Sump for Foundation Drain

Q. Should an exterior perimeter drain be connected to an interior drain and sump pit?

A. It's not a good idea to bring exterior water into the structure. Whenever possible, a "daylight" footing drain should be installed. If a daylight drain is not feasible, I'd recommend installing a sump pit on the exterior of the house. Locate the bottom of the pit well below the frost line (which is where the perimeter tile along the footing should be), and top it off with an insulated cover to prevent freezing (see illustration, opposite page). The pit should be at least 18 inches in diameter to allow room to install and service the pump. Since it's easier to install a sump and pump in the basement than in an outdoor pit, many builders do bring the exterior footing tile inside the foundation. If an exterior drain is connected to an interior sump, some form of backup pump should be installed. This can be a second line-voltage pump and a generator, or a battery-powered pump.

For houses on a municipal water supply, my favorite is a water-powered jet pump. The power can be off longer than a battery-powered pump will run, but municipal water pressure seldom fails. The jet pump

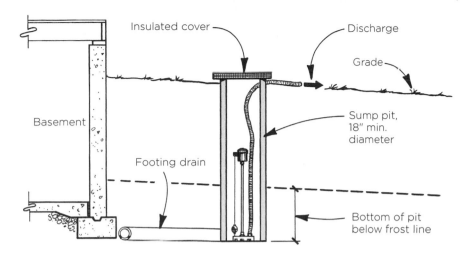

Insulated cover

Discharge

Grade

Basement

Sump pit,
18" min.
diameter

Footing drain

Bottom of pit
below frost line

If a sump pump is unavoidable, it should be installed in an outside sump pit with an insulated cover. Make sure the bottom of the pit is below the frost line.

has no moving parts, but it does use about as much water as it pumps and is limited to about 7 gallons per minute. This is not a serious limitation for a backup system, and it is less expensive than a battery-powered pump. One such pump is the Home Guard (Flint & Walling Pumps, Dean Bennett Supply, 800/621-4291; www.flintandwallingpumps.com)

Keep in mind that if the exterior footing drain is installed properly, there should be no need for an interior perimeter drain. If there is a spring under the floor or the water table is higher than the basement floor, a full basement should be avoided.   — *Henry Spies*

Best Foam for Foundation Insulation

Q. What kind of foam plastic insulation should be used on exterior foundation walls?

A. We have found that many different materials can work, but that some require more expensive installation than others to give equal performance. There's still much we don't know, but our investigation of 59 foundations in Minnesota and discussions with other experts in the field suggest the following:

- Extruded polystyrene (Amofoam, Certifoam, Foamular, and Styrofoam) always works well.
- Although only two specimens of *expanded* polystyrene (EPS) were observed in our study, we feel EPS can be used successfully. Also, a Canadian study found no performance problems with EPS below grade. However, we recommend only the higher-density products below grade, not the typical one-pound-per-cubic-foot variety found in most lumberyards.
- If you use polyisocyanurate or spray urethane, include a protective coating below grade that is long-lasting and strong enough to avoid puncture from backfill. In some cases, we observed higher water absorption, which may have been caused by damage to the below-grade protective coating.

A variety of products can work below grade, but extruded polystyrene consistently performs well in this application.

Dow Chemical

Also, foam plastic is not your only option. Consider high-density fiberglass or mineral wool board for sites where foundation-wall drainage is a must. *Editor's note:* Rigid fiberglass panels are available as commercial roof insulation from Owens Corning (www.owenscorning.com), and rigid mineral wool DrainBoard panels are available from Roxul Inc. (www.roxul.com).

No matter what type of material you use, two elements are essential to prevent moisture absorption and deterioration. First, insulate to the top of the foundation wall, making sure to leave no gap where the insulation meets the siding. Use mechanical fasteners or adhesive to prevent the insulation from slipping down the wall.

Second, protect the insulation above grade from physical abuse and sunlight. Since many of the installations we have seen have missing or damaged coatings, we recommend an above-grade covering material at least as durable as the siding. Examples include 1/2-inch pressure-treated plywood, high-quality stucco, and fiberglass or fiber-cement panels.   — *Bruce Nelson*

Insulating a Foundation

Q. To turn a small outbuilding into a living space, I plan to pour a slab floor within the existing 4-foot frost wall foundation. How much and what kind of insulation should I use for the foundation, and where should I place it — under the slab or next to the frost wall?

A. Assuming you wish to use exterior insulation, I suggest that you insulate the exterior surface of the foundation wall using extruded polystyrene foam boards. You can place R-5 rigid insulation (1-inch board) to the full 4-foot depth, or use R-10 rigid insulation (2-inch board) to a depth of 2 feet. These two options have almost the same thermal performance. One advantage of the full-depth treatment is that the insulation board will be supported by the footing. Insulating under the floor is generally not cost effective unless you are using radiant heating in the slab. Either choice of half-height or full-height exterior insulation will save energy and cut heating bills. For a small building in New England, for example, a simple calculation (using various assumptions about energy costs and heating system performance) shows that you can expect savings ranging from $50 to $100 per year from that level of insulation compared with an uninsulated foundation. Interior insulation with foam or studs and fiberglass batts is another option that will yield similar performance.
— *Moncef Krarti*

Crawlspace Insulation

Q. I am building an addition over a 30-inch-high crawlspace. The local inspector has approved my plan to build a sealed crawlspace without ventilation. Should I install rigid foam insulation on the interior of the block walls, or should I insulate between the floor joists?

A. The International Residential Code permits unvented crawlspaces under certain conditions — for example, if the crawlspace has a ground cover, is properly insulated, and is mechanically ventilated. We recommend insulating the foundation walls rather than the floor. Insulation in a floor is frequently interrupted by wires, pipes, and bridging, making it next to impossible to get a tight air seal.

The first step to detailing a sealed crawlspace is installing good drainage when the foundation is built. Managing bulk water is essential for success in a sealed crawlspace. I like to see the perimeter drain at the level of the footing, sloped to one corner of the building. An exterior footing drain is the minimum; even better is to have drains on both sides of the footing, connected across the footing at the low corner (see illustration).

The crawlspace floor should be sloped to that low corner, so that if a water pipe breaks and floods the space, water will drain that way on the poly. To provide the water with an escape route, we install a backwater valve in that low corner. This valve is connected to the perimeter drains; when we later install the final poly, we cover the inlet with a grate that's sealed to the poly with mastic. When the crawlspace is dry, the valve stays closed and keeps the groundwater out. But if water collects in the corner of the crawlspace, the valve flapper opens and allows the water to drain out through the footing drain.

Codes usually require a foundation to be damp-proofed before backfilling if the exterior grade is higher than the interior crawlspace floor. Damp-proofing will reduce the leakage but won't prevent it, so I say use the belt *and* the suspenders: Coat the foundation all the way to the footing, and install drains inside and out.

Before installing the insulation, we first attach 6-mil black plastic to the wall with a water-based adhesive mastic. Where pipes or wires go through the side wall, we have to fit the foam board to them carefully and seal the vapor barrier around them. We use a water-based, nontoxic, duct-sealing mastic called PS-1, from RCD Corporation (800/854-7494, www.rcdcorp.com), for all our sealing work. It sticks to all kinds of materials, rough or smooth, and even to dirty surfaces. The wall poly extends onto the ground about a foot, leaving an edge flap for us to seal the floor poly to later. We then fasten 2-inch foam board (either foil-faced polyisocyanurate foam or extruded polystyrene) over the poly with powder-driven masonry nails.

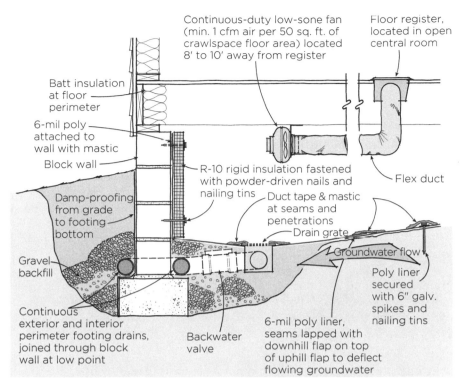

Continuous-duty low-sone fan (min. 1 cfm air per 50 sq. ft. of crawlspace floor area) located 8' to 10' away from register

Floor register, located in open central room

Batt insulation at floor perimeter

6-mil poly attached to wall with mastic

Block wall

R-10 rigid insulation fastened with powder-driven nails and nailing tins

Flex duct

Damp-proofing from grade to footing bottom

Duct tape & mastic at seams and penetrations

Drain grate

Groundwater flow

Gravel backfill

Poly liner secured with 6" galv. spikes and nailing tins

Continuous exterior and interior perimeter footing drains, joined through block wall at low point

Backwater valve

6-mil poly liner, seams lapped with downhill flap on top of uphill flap to deflect flowing groundwater

The components of a sealed crawlspace must address the issues of bulk water, air sealing, vapor sealing, and thermal insulation. Damp-proofing and perimeter drains are necessary details, because the poly ground cover is not designed to handle constant water contact. Drains, both inside and outside the footing, joined at the low point, are preferable. Tie a floor grate into the interior footing drain and protect the link with a backflow preventer valve. Floor poly is lapped reverse-shingle fashion so as to manage water flowing in the soil beneath the plastic. For maximum resistance to air and vapor infiltration, seams in the poly and at penetrations are sealed with mastic. R-10 insulation is placed at the foundation wall, aligning the air, vapor, and thermal boundaries of the space in the same plane. Finally, the fan ventilates the space with conditioned house air.

During construction I recommend installing temporary poly on the floor of the crawlspace. This will keep moisture out of the structure as it's dried in, and all the trades can do what they have to do, which will inevitably tear up the poly a bit. A week or two before the house is ready for the owners to move in, we take out the temporary poly and put down a nice, clean, 6-mil poly floor

Each joint in the new poly is secured with duct tape that is then coated with a brushed-on layer of mastic. The mastic should extend at least an inch beyond the tape on each side.

Take note: it matters how you lap the poly at joints. You need to create a shingle effect to handle flowing water, but you're shingling upside down. If you lap the joint one way, it will allow the water *underneath* to flow past, but if you lap it the other way, the lap will scoop water and hold it against the seam, threatening your mastic seal. To protect the permanent poly against future wear and tear, we also lay carpet runners in the space as a path for service technicians, running from the access door to the water heater, furnace, and any other appliance.

At this point, the crawlspace is clean, dry, well insulated, and both airtight and vapor-tight. We recommend it also be actively conditioned with a dedicated, low-sone fan rated for continuous duty. The steady supply of conditioned air from this fan will maintain the crawlspace at a humidity close to that of the main house. We typically have it installed under the floor about 8 or 10 feet away from a floor register, connected by a run of flex. Separating the register and the fan in this way makes for quieter operation. — *Jeff Tooley*

Insulating Walk-Out Basements

Q. What is the best way to insulate the footing, foundation, and slab on the walk-out side of a basement?

A. Probably the easiest way is to pour the footing and insulate the inside of the stem wall with 2-inch-thick extruded polystyrene foam (see illustration). The insulation should extend at least 2 feet below grade, and the top can be tapered to a thin edge so it is concealed by the interior trim. This works particularly well if there are heating ducts in the slab near the perimeter.

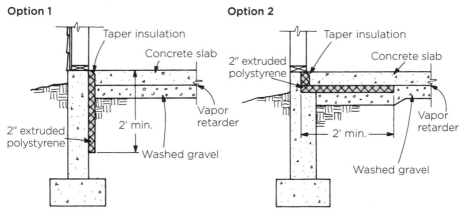

Option 1 · Option 2

Taper insulation
Concrete slab
2″ extruded polystyrene
Vapor retarder
2′ min.
2″ extruded polystyrene
Washed gravel

Taper insulation
Concrete slab
2″ extruded polystyrene
Vapor retarder
2′ min.
Washed gravel

To insulate the walk-out side of a basement, install 2-inch-thick extruded polystyrene on the inside of the stem wall after the footing has been poured (Option 1). An alternative is to insulate the slab perimeter at least 2 feet in from the outside edge (Option 2). In either case, the top edge of insulation can be tapered to a thin edge so it is easily concealed by trim.

An alternative is to insulate the slab perimeter, extending the insulation at least 2 feet from the edge. Again, where the insulation turns up along the perimeter, it should be tapered so it can be concealed.

A third option, of course, is to insulate the outside of the foundation with extruded polystyrene. The portion above grade should be protected with stucco, fiber-cement paneling, or foundation-grade pressure-treated plywood. However, this system should not be used if the walls have a masonry veneer. — *Henry Spies*

Waterproofing ICFs

Q. Is there a spray-on waterproofing that's compatible with ICF foundations — that won't "melt" the polystyrene?

A. When waterproofing ICFs, we use a spray-on polymer-modified membrane — either Tremco Barrier Solutions' Tuff-N-Dri (800/876-5624, www.tuff-n-dri.com) or GMX's Ultra-Shield (800/762-8225, www.gmxco.com). The GMX product comes with a 20-year warranty, even when applied to ICFs. The key thing with ICFs is to use a water-based formulation rather than a solvent-based one, which would destroy the polystyrene. Even though these coatings have the ability to withstand hydrostatic pressure, we always install a vertically stranded drainage board to protect the membrane during backfill. With ICFs, the R-value of the drainage board is not important, so the minimum thickness (typically 3/4 inch) can be used. As with any foundation drainage method, a perimeter drain tile in a bed of clean stone is also needed; usually, the coating manufacturer's warranty requires it. — *Bruce Richgels*

Finishing Basement Walls

Q. To finish an existing basement, I will be installing 2x4 walls around the perimeter. Should I install some type of waterproofing to the concrete walls? Can the 2x4 walls touch the concrete, or should I leave an air space?

A. Before finishing the interior of a basement, you must verify that the basement doesn't leak, and that the exterior of the foundation is protected by damp-proofing, good drainage, and controlled surface runoff.

I believe that applying an additional layer of damp-proofing on the interior surface of the foundation wall makes sense, if only as relatively cheap insurance. I have had good luck using Sto Watertight Coat (Sto

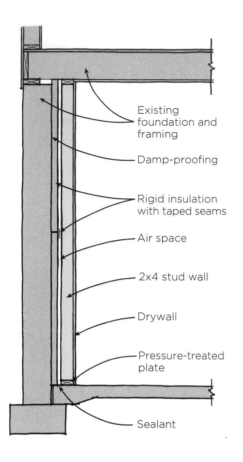

Existing foundation and framing

Damp-proofing

Rigid insulation with taped seams

Air space

2x4 stud wall

Drywall

Pressure-treated plate

Sealant

Corp., 800/221-2397; www.sto-corp.com), a two-component, trowelable, cementitious compound that has a low-perm rating.

Your 2x4 walls should have pressure-treated bottom plates and should be spaced away from the foundation. Most foundation walls are not perfectly plumb and straight, so it is easier to keep your wall surface true if you space the frame away from the concrete wall. Also, building codes prevent you from placing nontreated wood in contact with the foundation.

Air-sealing a wall can be difficult, so after installing interior damp-proofing, install rigid foam insulation directly to the inside surface of the foundation walls, using construction adhesive. Caulk and/or tape the seams of the rigid foam to make it airtight, so that warm interior air can't reach the cold foundation. Then build an uninsulated wood-frame wall that is spaced away from the foundation. This way, it's easier to run plumbing and wiring, and the wall should remain above dew point temperature, reducing the likelihood that condensation will form in the wall.

— *Paul Fisette*

Leaving Rigid Foam Exposed

Q. I plan to install rigid foam insulation in a crawlspace. As far as I know, most types of rigid foam insulation can't be left exposed, but rather need to be covered with a layer of drywall for fire resistance. Is there a type of rigid foam insulation panel available that can be left exposed?

A. Not that I know of. However, most building codes allow the installation of unprotected rigid foam in a crawlspace that has no open connection to a basement, as long as the foam has passed alternative testing procedures. Dow's Styrofoam (866/583-2583, www.dow.com) and Owens Corning's Foamular (800/438-7465, www.owenscorning.com) are two products that pass the burn test and can be left unprotected in an isolated crawlspace. Building codes restrict the use of unprotected foam in habitable or accessible spaces. Different code jurisdictions may have particular code sections that deal with this issue. Some codes say that if your crawlspace is connected to a basement, you need to cover the foam with an ignition barrier like 1/4-inch plywood or particleboard. Where inspectors draw the line often depends on whether or not there is a mechanical system in the crawlspace. The thinking here is that a fire could be set off by either the equipment or activity of people in the space. When in doubt, it always pays to talk to your inspector during the design stage. — *Paul Fisette*

Retrofitting a Window Into a Block Foundation

Q. I need to install a basement egress window in a concrete-block foundation wall. What kind of header should I use to support the eaves wall above the new window opening, and how should I install it?

A. After you've double-checked the dimensions of your window to verify that it meets egress window code, I'd recommend that you also double-check the conditions around the proposed opening in the foundation. If excavation will be required to accommodate the new window, be aware that making grade changes could introduce complications. You don't want water running toward the house, so you may need to add money to the budget to address that issue.

You'll also need to temporarily support loads near the opening in the basement with a 4-by-8-inch header that's long enough to hold up the joists above the window opening plus two extra joists on each side. This header should be temporarily supported on posts. Use 2-foot-long blocks under each post to spread the weight over the basement slab. Install the temporary header close to the wall, but leave enough room to work. While you can build the permanent header out of concrete-filled lintel blocks supported by angle irons, an easier option is to use a precast reinforced-concrete lintel. These are available at most building yards, and will need to be sized to your specific opening and cut to length on site. Because these lintels may develop cracks over time, I like to reinforce them with angle iron as added insurance against callbacks (see illustration, next page).

The first course of block below the mudsill will house the lintel, which should extend roughly 6 inches in either direction past the window's rough opening. Using a level, mark the lintel and window roughs

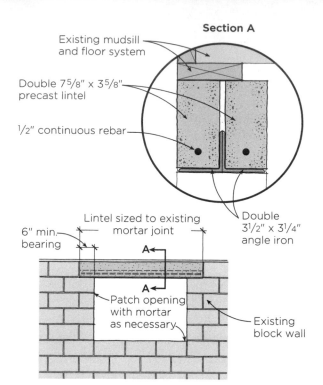

Section A

Existing mudsill and floor system

Double 7⅝" x 3⅝" precast lintel

½" continuous rebar

Double 3½" x 3¼" angle iron

Lintel sized to existing mortar joint

6" min. bearing

A

A

Patch opening with mortar as necessary

Existing block wall

When cutting an opening in an existing block wall, try to align it with existing mortar joints, and size the new lintel so that it is supported by at least 6 inches of block on either side. Though structurally unnecessary, a pair of angle irons provides extra support that helps eliminate minor cracking in doubled-up precast lintels (Section A).

on the foundation wall, aligned if possible with existing mortar joints. The outline will resemble a rather fat T. Using a masonry drill held as level as possible, drill through the wall at all six corners to mark your roughs on the exterior wall of the foundation. Use a circular saw or a grinder equipped with a masonry or diamond blade to score the wall on the interior and exterior markings. This will create plenty of dust, so be sure to protect yourself and the room.

Once the wall has been scored, start to chip away, starting in the center of the opening and working toward the perimeter. We use a rotary hammer but always have a three-pound lump hammer and cold chisel close at hand; we periodically remove the debris underfoot to keep the work area clean. Once you've opened up the wall, you'll have a number of partial concrete blocks that will need to be filled or replaced. Repair the opening as required, adding wire mesh to the cavities to give your mortar something to hold on to.

Finally, install the lintel, shimming or jacking it flush to the sill and packing the bearing shelves you've created underneath with mortar. Once the lintel is secure and the mortar has set, you can go ahead and remove the temporary header and posts, apply stucco as required, and install the window.

— *Rob Corbo*

Preventing Cracks in Concrete Slabs

Q. What causes a concrete slab to crack, and how can it be prevented?

A. Concrete cracking is caused by stresses resulting from drying shrinkage, temperature change, or applied loads.

Drying shrinkage is an inherent and unavoidable property of concrete. During the setting and hardening stages, excess mixing water in the concrete evaporates, causing the concrete to dry from the surface inward. Shrinkage begins near the surface, pulling at the moist inner portions of the concrete, which are restrained by friction on the subgrade, by reinforcing steel, and by building connections. This restraint prevents the concrete from shrinking freely and uniformly, resulting in cracking.

While drying shrinkage and some cracking is inevitable, it can be reduced by specifying adequate compressive strength, minimizing the water content, spacing control joints properly, and adequately curing the concrete.

Compressive strengths are governed by local building codes. In general, basement walls require a minimum 2,500 psi concrete, while flatwork ranges from 3,000 to 4,000 psi. For residential work, recommended slumps range from 3 to 5 inches for flatwork, and 5 to 7 inches for basement walls. Once the concrete is ordered to a specified slump, don't add more water at the site to speed the pour.

The purpose of control joints is to confine cracking to predetermined points in a slab, rather than letting them occur randomly. Control joints should be tooled or sawn to a depth of one-quarter the slab thickness. Joints should be spaced at intervals not more than 30 times the slab thickness. Driveways wider than 10 feet require both transverse and longitudinal control joints.

Curing helps reduce shrinkage cracking and maintains slab strength. Typically, curing involves keeping the concrete moist and covered for five to seven days, or applying a spray-on compound that forms a membrane on the surface.

Extreme temperature changes that occur anytime from immediately after to a year after slab placement can have the same adverse effects on concrete as drying shrinkage. Proper control-joint spacing is the most effective method to guard against this.

Applied-load, or load-stress, cracking occurs when the weight of an object on a slab stresses the concrete beyond its tensile strength. Such cracking often occurs, for example, when a heavy truck drives over a

sidewalk designed only for pedestrian and light vehicular traffic. To prevent load-stress cracking, make sure a slab is built over a uniformly compacted, well-drained subgrade, and is thick enough to withstand the kind
of use it will get. In residential concrete, 4 inches is the minimum thickness for walkways and patios. Garage slabs and driveways should be 5 to
6 inches thick if any heavy truck traffic is anticipated, otherwise 4 inches
is adequate. — *Don Marsh*

Fiber vs. Wire Mesh Reinforcing

Q. Which is better at preventing cracks in concrete
slabs: fiber reinforcement or wire mesh?

A. Adding synthetic fibers, such as Fibermesh, to a concrete mix can
help reduce plastic shrinkage cracking while the concrete is setting,
whereas wire mesh can help prevent shrinkage cracking over the next
several weeks while the concrete is drying. However, neither fiber nor
wire mesh will prevent all cracking. In fact, a concrete slab will always
crack as it shrinks during curing. By placing control joints at the proper
spacing, you can limit where those cracks occur (see table of recommended joint spacing). If, however, a crack occurs where you did not
expect it, wire mesh is better at keeping the crack tight and preventing it
from opening up further. — *Robert Shuldes*

Maximum Spacing of Slab Control Joints

Slab thickness	Slump 4 to 6 in.		Slump less than 4 in.
	Aggregate less than 3/4 in.	Aggregate 3/4 in. and larger	Aggregate 3/4 in. and larger
4 in.	8 ft.	10 ft.	12 ft.
5 in.	10 ft.	13 ft.	15 ft.
6 in.	12 ft.	15 ft.	18 ft.

Source: Portland Cement Association

Supporting New Slab Next to Existing Slab

Q. What's the best way to support a new garage slab next to an existing foundation wall?

A. There are many ways to support a garage slab next to a basement wall. The best approach is to provide compacted backfill to 95% Proctor density from the footing to the sub-base. The sub-base should consist of 4 to 6 inches of coarse aggregate (Illustration A). If the job schedule or budget does not allow for careful soil compaction around the basement wall, then any area of over-excavation should be backfilled with minimum $1^1/2$-inch-diameter clean gravel, which will self-compact evenly under the weight of the slab and any future loads.

Another option is to use a shelf angle to support the edge of the garage slab (Illustration B). We usually recommend a continuous $3x3x^3/8$-inch steel angle, bolted to the concrete foundation wall every 16 inches with $1/2$-inch-diameter bolts. However, this will support only the slab edge and will not prevent stress cracking that may occur as the soil under the slab consolidates, leaving voids where the slab is not supported. A preventive measure is to use #3 or #4 reinforcing steel placed 12 inches on-center in both directions in the slab. This will enhance the slab's ability to span the voids as soil consolidation occurs.

Slabs can also be connected to the basement wall with rebar pins (Illustration C). This is essentially the same concept as a shelf angle. However, because the steel reinforcement restricts slab edge movement,

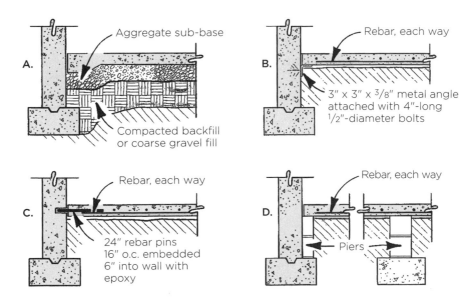

A. Aggregate sub-base

Compacted backfill or coarse gravel fill

B. Rebar, each way

3" x 3" x 3/8" metal angle attached with 4"-long 1/2"-diameter bolts

C. Rebar, each way

24" rebar pins 16" o.c. embedded 6" into wall with epoxy

D. Rebar, each way

Piers

there may be more cracking on the garage floor.

Another technique is to support the slab on wall pilasters or concrete piers (Illustration D) that bear either on the footing below or wherever areas of soil subsidence are suspected. — *Brent Anderson*

Reinforcing a Slab Over Disturbed Soil

Q. I need to pour a thickened-edge slab foundation for a two-story garage apartment across a recently dug 2-foot-wide utility trench. The backfill over the trench has subsided, and I'm concerned that the foundation may sag and crack at that point. Can I add rebar to the slab to span this area, or is it better to mechanically compact the trench area?

A. If you're in doubt but don't plan to have an engineer's compaction test, you should probably do both. Run over the trench area with a plate compactor, preferably, or at minimum a jumping jack, and top up any depression that's left with the clean stone that you spread under the slab. Then, for added insurance, add rebar to the slab where it crosses the trench area. Use 1/2-inch or 5/8-inch rod, placed perpendicular across the trench and spaced 12 inches on-center. Cut the rebar so you have at least 1 to 2 feet of length over the undisturbed soil on both sides of the utility trench. — *Jay Meunier*

Shallow Foundation for Flat Site

Q. I have a client who wants me to build a 30x40-foot unheated barn/garage with a slab floor. The area where the building will sit is very flat, with no lower terrain nearby, so it's not feasible to build a typical 4-foot stem wall with a daylight drain. Frost depth is 48 inches around here (northern Vermont). I wonder if a shallow frost-protected slab would work in this application. Do I need to thicken the edges? And do I need a rebar grid in the slab?

A. You have an ideal application for a frost-protected shallow foundation. The method has been approved for use in the 2003 *International Building Code*. The design guide for heated structures is published right in the code. For unheated structures, the code references ASCE32-01, and

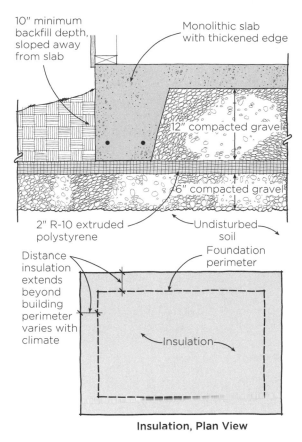

10" minimum backfill depth, sloped away from slab

Monolithic slab with thickened edge

12" compacted gravel

6" compacted gravel

2" R-10 extruded polystyrene

Undisturbed soil

Foundation perimeter

Distance insulation extends beyond building perimeter varies with climate

Insulation

Insulation, Plan View

By combining rigid foam insulation with a well-compacted, free-draining substrate, a frost-protected shallow foundation prevents heaving damage, even in an unheated building in the coldest climates.

that's where you have to look to find the design standards. At your location in Vermont, you have an air-freezing index of 2,000 freezing degree-days, and a mean annual temperature of 45°F. According to table A8 in ASCE32-01, your ground insulation should have an R-value of 10. It should extend 60 inches beyond the building perimeter and be at least 10 inches below finished grade (see illustration). You should place this insulation on 6 inches of non-frost-susceptible fill (sand or gravel).

You also need to check the building loads to make sure you don't exceed the bearing capacity of the rigid foam. This tells you how wide the base of your thickened footing must be. Extruded foam boards are available in a variety of compressive strengths: 15-psi, 25-psi, 40-psi, and more. Some products (like Pactiv GreenGuard, 800/241-4402; www.greenguard.com) come standard with 25-psi compressive strength. This gives 3,600 psf, typically strong enough unless you have a heavy point load. Other foams, like Dow Styrofoam (866/583-2583, www.dow.com/styrofoam) and Owens Corning Foamular (800/438-7465, www.owenscorning.com), commonly come in both 15 and 25 psi, so you need to watch what you're buying. If you need a stronger foam, you can get it as a special order. I usually use a factor of safety of 2.5 to 3.0 on the foam loading to account for long-term creep potential.

To build the foundation, first prepare your subgrade pad 24 inches below the top of your finished slab. Remove vegetation and loose dirt and either cut the high spots or use compacted gravel fill to level the pad.

Extend this prepared area 5 feet beyond your building perimeter. Next, place an additional 6 inches of compacted, non-frost-susceptible gravel fill. Then lay 2 inches of rigid foam insulation over the gravel, under the entire building area, extending the foam board 5 feet beyond the building perimeter. Now form the slab perimeter. The code requires a minimum depth of 10 inches of dirt cover over the foam.

So assuming you want your slab to be 6 inches above finished grade, the thickened slab edge will be 16 inches tall. I usually use a 14-inch- to 15-inch-tall edge form. This is a little shorter than the required 16-inch slab edge, but it allows for some irregularities in the subgrade. Drive your form stakes right through the foam. Then place 12 inches of compacted gravel fill on top of the foam inside the building and shovel out the thickened edge around the perimeter. Install rerods, and you're ready to pour. I would definitely use two 1/2-inch rerods in the thickened edge. The grid in the slab provides extra insurance if the budget allows, but it is not necessary if the base is properly prepared and compacted.

Pour the slab, strip the forms, backfill the outside with 10 inches of dirt, and you're done. *— Bill Eich*

Garage Slab Drainage

Q. How should a garage slab be sloped for drainage toward the overhead door? At 1/4 inch per foot on a 24x26-foot foundation, the height difference from front to back would be more than 6 inches. Do I need to grade the sand and gravel sub-base at the same slope so the slab can have a uniform 4-inch thickness?

A. A garage floor does require a slope for drainage so water won't pool on the floor, and 1/4 inch in a foot is typical. But if you do a good tight screed and finish job, 3/16 inch per foot will do. And yes, the gravel sub-base should be graded to the same slope as you intend for the slab surface. The bigger question is, where do you drain the water to? If you are located in a freeze-thaw climate such as we have here in South Bend, drainage toward the overhead door could be a problem.

When snowmelt or water from washing the car seeps under the door and meets 25°F or colder temperatures, it freezes. Frequent seepage can repeatedly melt and refreeze the existing ice, setting up a continual freeze-thaw cycle that can cause the surface to scale off early in the driveway's service life. One solution is to pitch the slab to the center of the garage and provide drainage into a dry well or, if the elevation allows, to an exterior drain. But even with drainage through the overhead doorway, many concrete driveways and aprons hold up well under freeze-thaw conditions. The keys are to make sure you have air entrainment in the mix, seal the slab and driveway when new, and then instruct the home-owner to maintain it properly.

We seal new driveways and slabs with Kure-N-Seal 30 from Degussa Building Systems (800/433-9517, www.chemrex.com). The frequency of resealing depends on wear patterns for a particular driveway. Basically, as soon as the sealer has worn off, it's time to apply a new coat. A good test is simply to wait until the drive is perfectly dry and apply a few separate drops of water on the drive using an eyedropper. If the drops soak in, you need to seal the surface; if they bead up, you're okay.

If you do install the drain in the center of the garage, be sure to check your local and state building codes. Some may allow a dry well, while others may require connecting to an exterior drain. — *Rocky Geans*

Spalling Concrete Floors

Q. Here in Minnesota, many garage floors suffer from spalling. What causes spalling concrete, and is there any cure?

A. I am not aware of any cure for spalling slabs, but there are ways to prevent it. Spalling can be caused by poor weather conditions during finishing, improper finishing techniques, or damage from de-icing salts. Spalling slabs are more common in climates with frequent freeze-thaw cycles. Such climates require the use of air-entrained concrete (concrete with small bubbles of air to provide room for expansion when moisture freezes).

Assuming that air-entrained concrete was used, most spalling problems can be traced to errors during the pour. Either the slab dried too quickly, or the concrete slab was worked too early, trapping the bleed water just below the surface. When a slab is poured on a hot, windy day, it can dry too quickly. When that happens, some workers apply small amounts of water to the surface of the concrete to get the cream back to a workable condition. This reduces the strength of the concrete surface. The weaker surface layer can later spall when moisture freezes and pops the cream. If the slab is worked too early, bleed water can be trapped in the concrete, since the concrete just below the cream on the surface has not yet set up. A weak layer in the concrete will form where the bleed water was trapped. Again, when the slab later freezes, the concrete can pop along the line at the weak layer.

Choose a concrete contractor with a reputation for quality work. Don't pour a slab that is open to the weather on a hot, sunny, windy day. Finally, always apply a curing-sealing compound to allow the slab to cure more slowly and seal the top from contaminants like road salt. — *Jay Meunier*

Demo Near Post-Tension Cables

Q. What precautions should you take when tearing out a post-tensioned slab? What are the risks? We are pouring a large post-tensioned slab, and while there is no demo involved at this stage, I would like to know how to proceed in case there is a plumbing mistake that requires us to cut out and repour a section of the slab.

A. I installed dozens of post-tensioned slabs in homes in Las Vegas, as well as a 10,000-cubic-yard post-tensioned parking garage. The cables are tensioned to approximately 30,000 pounds, and if you cut one the wrong way, the force can easily kill you. I've seen cut cables rip out of slabs on several occasions. One ripped open a 10-foot-long strip of slab and sliced through a shear wall. Another slammed into a dishwasher 5 feet from the hole and sliced it in half. In both cases, the geniuses who caused the damage were lucky enough to avoid harm. In most cases, the cables can be destressed by a post-tension contractor. It's not an easy task, especially if the location of the cables isn't easy to establish, but it's the only way to proceed unless you plan to do all the demo from the cab of a very large excavator with a hydraulic hammer and can keep everyone far from the area.

I wouldn't recommend that avenue, though. Here's what I'd do in your case: Just before the pour, take photos of every slab penetration that could possibly be in the wrong location, marking dimensions to the nearest cable in each direction. Note this information on a drawing as backup in case the pictures aren't clear. Also, spray-paint the location of the "dead" ends of the cables on the formwork. (These are the ends that, rather than running all the way to the slab edge, get embedded in the middle of the slab.) Transfer these marks to the slab before stripping the forms, so you'll be able to locate the dead ends if you need to. Mark the "live" end locations on the top of the slab before cutting the cables and patching the holes. Assuming the cables run relatively straight, this will allow you to "connect the dots" and roughly locate a cable in the slab. We did this recently on the post-tensioned garage slab, spraying paint along the cables on the deck forms; it really helped when we had to recore a few plumbing penetrations. Snap out all your walls before tensioning the cables, and determine what, if anything, needs to be moved. Jack up the slab now and make the moves so that you don't hit any stressed cables: Repairing an unstressed cable is far easier than repairing a stressed one. Usually you end up waiting at least a week or so after the pour to tension, so this shouldn't be a big deal. If it's not going to cause a tripping hazard, leave the ends of the tendons hanging out for a while so that you can destress a cable later if you have to. Unfortunately, it's not usually possible to leave cables hanging because people end up tripping on them and getting grease on themselves.

— *Bob Kovacs*

Preventing Frost Action on Structural Posts

Q. In New York state, we built a pergola (a type of garden trellis) using 8x8 pressure-treated posts set 4 feet into the ground. During the first winter, the frost lifted the posts. What should we have done to keep the posts from heaving?

A. To resist wind uplift, pergola posts should be securely anchored to the ground. For a solid anchor, I prefer to embed pergola posts in concrete, by inserting each post into a Sonotube. If an irregular-shaped hole is used instead, frost will tend to grip the concrete and heave the post. (Another option is to extend the Sonotubes a little above grade and to use embedded post anchors with sufficient hold-down strength).

Check with your local building officials for information regarding the frost depth in your area. Your post holes should be dug to the frost depth plus 6 inches. Each hole should receive 6 inches of crushed stone

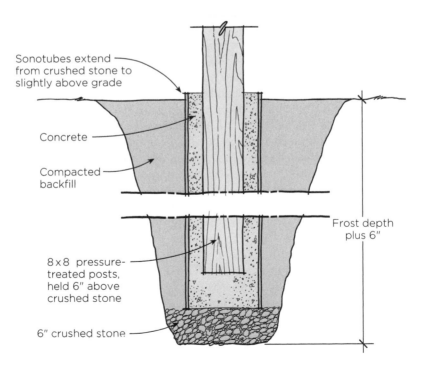

Sonotubes extend from crushed stone to slightly above grade

Concrete

Compacted backfill

8 x 8 pressure-treated posts, held 6" above crushed stone

6" crushed stone

Frost depth plus 6"

for drainage. Drainage is important, since frost heaving is more likely in wet soil than dry soil.

The Sonotubes should extend from the crushed stone base to slightly above grade. After backfilling around the Sonotube with compacted dirt, insert the pressure-treated posts inside the Sonotube, holding the bottom of the posts about 6 inches above the crushed stone in order to provide a space where concrete can flow under the posts. Then plumb and brace the posts and fill the Sonotubes with concrete.   — *Ron Hamilton*

Sinking Foundation

Q. I'm looking at a wood frame house with a block foundation that is settling. The basement wall sits on a 20-inch-wide concrete footing, but someone later poured a concrete shelf from footing to grade and then put brick veneer siding on the house. The soils can't carry the extra weight. What can I do to give the house sufficient support?

A. That house needs a deep pier system to support the footings from below. It's a job for a specialty contractor. Your choices on the market are concrete pilings, push piers, or helical piers; which one you choose will depend on what the local contractors in your area have available and on which contractor you feel most confident about. I'd get at least three proposals before you decide.

Some systems can be installed from inside the basement if there are exterior elements you don't want to disturb. But, generally, you end up excavating around the exterior down to the footing and attaching your piers to the footing from the outside. Concrete pilings are probably the costliest way to go. That involves auguring a 24-inch hole wherever you need a pier, placing a rebar cage in the hole, and pouring a concrete pier. To raise the house, you have to terminate the pier 2 feet below the footing, with a shoulder that extends under the footing where you can place your hydraulic jacks. If you're just trying to stabilize the foundation but not lift it, you bring the shoulder or shelf right up to the bottom of the existing footing.

Push piers, your second choice, are steel tubes that are pushed down into the soil with a hydraulic rig. The depth is determined by the resistance that builds up as the pier is pushed deeper. The piers attach to the footings with metal brackets. Again, if you're lifting, you'll have to provide for the jacks. The option I prefer is helical screw piers. These are screwed or augured into the soil rather than driven. If you have bedrock a reasonable distance down, I'd use just one helix per pier and take it down to rock. But if your rock is too deep, you have to rely on the soil to carry the load. In that case, you add more helixes. With two or three helixes per pier, you may have to go down only 8 or 10 feet below the footing, depending on the soil.

The soil is a critical factor — its bearing capacity will determine pier spacing and depth. It's well worth the money to have a soils engineer evaluate the job, do a soils test, and place the piers for you. The cost is per pier, and you're likely to need a pier every 4 to 6 feet wherever the footing needs support. — *Dave Cunningham*

Protecting Below-Grade Framing

Q. We are building an addition and want the floor levels between the old and new living areas to line up. However, this would put the floor framing slightly below the existing grade. Is there a foundation flashing detail to protect the floor framing and siding in this case?

A. This is a common problem when adding on to a house, but flashing is not the answer. The International Residential Code requires the use of pressure-treated lumber for all wood framing that rests on a concrete foundation and is less than 8 inches above finished grade. Steer away

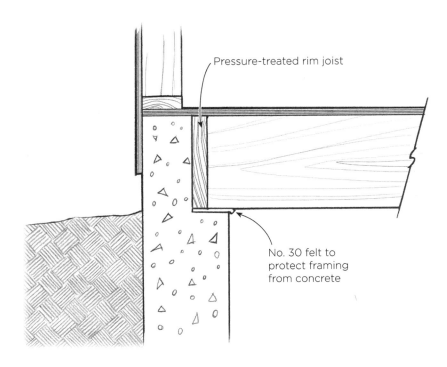

Pressure-treated rim joist

No. 30 felt to protect framing from concrete

from looking for a flashing treatment. You want to fix the problem, not treat the symptom.

First, determine whether the existing grade can be altered to eliminate the problem. Excavating for an addition foundation usually disrupts the surrounding grade, anyway, so re-sculpturing the site to accommodate the addition may not necessarily be that complex or expensive. The "cut" required to lower the grade a foot or so can often be incorporated with landscaping or low retaining walls.

If you're faced with a tight lot, of course, adjusting the level of the grade may not be an option. In this case, you can extend the foundation upward to cover the rim floor framing, as shown in the illustration (previous page). Be sure to use treated lumber in places that come in contact with the concrete, and use a capillary break to separate the framing from the concrete (No. 30 roofing felt works well).

This detail has its drawbacks. Gaining access to the wall cavity from below (for duct runs and plumbing, for example) is nearly impossible. Be sure to have your mechanicals worked out well in advance if you plan to use this approach. Also, forming the shelf in the foundation can be costly, especially if the extra height must be added to standard 8-foot concrete forms.

— *Carl Hagstrom*

Skirts for Pier Foundations

Q. What's the best way to build a skirt around an addition that's built on piers or posts?

A. Our company has built many additions on wooden posts and concrete piers, and we've developed a simple technique to close off the open area between the floor system and the ground. We install a pressure-treated plywood skirt that matches the thickness of the wall sheathing above (see illustration). The top edge of the plywood skirt is nailed to the rim joist; the lower portion extends about 1 foot below grade and is fastened to a treated 2x4 nailer that is held a few inches above the ground. The plywood can be

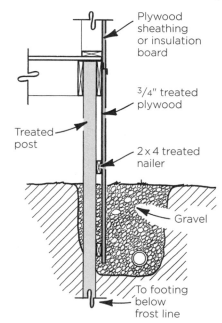

Plywood sheathing or insulation board

3/4" treated plywood

Treated post

2 x 4 treated nailer

Gravel

To footing below frost line

This simple skirt works well for room additions on wood posts or concrete piers. In heavy soils, good drainage prevents frost heaves.

sided to match the walls, or parged (over wire lath) to simulate the look of a masonry foundation.

We've never had a problem using this approach in our area of Indiana, where the frost line is 3 feet and local soils drain well. In colder regions with heavy clay soils, it would be wise to surround the buried portion of the plywood skirt with gravel, and place a drain to daylight in the bottom of the trench.

No matter what your local conditions are, check in with your building official for approval of this post-and-skirt foundation system before breaking ground. — *David Bowyer*

Is Rusty Rebar Okay?

Q. All of the foundation specs we build on require rebar to be free of rust and mill scale. For years this hasn't been an issue, but recently a project manager called us on it. All we can think to do is wire brush the entire lot of rebar. Is this really necessary?

A. Fortunately, there are a couple of standards you can cite in your defense. The ASTM standard for deformed steel reinforcement (A706) and the Concrete Reinforcement Steel Institute (CRSI) *Manual of Standard Practice* both give the same recommendation: Reinforcing bar with rust, mill scale, or a combination of both is satisfactory, provided the minimum dimensions, weight, and height of the deformations (the formed protrusions on the bar) of a hand-wire-brushed test sample are not less than the values specified in the standard. In other words, if the rust or mill scale is light, it will not affect the bond to the concrete. In fact, studies have shown that mill scale and light rust enhance the bond between concrete and steel. — *Tim Fisher*

How Soon Can Forms Be Stripped?

Q. How soon can concrete forms be stripped from a foundation wall? Can we build on them as soon as they're stripped?

A. The forms should remain in place for at least 8 to 12 hours after the wall is poured if the temperature is above 50°F. You can begin laying your sills and joists and building light frame walls immediately thereafter. But avoid backfilling or other work that places a lot of lateral stress on the foundation walls for at least seven days after the pour. Also, do not back-fill until the foundation walls are braced or the first floor deck is in place. After one week, the concrete should reach 80% of its ultimate strength. If the concrete is placed in cold weather, you should wait longer before building and backfilling in order to compensate for the delay in the set-ting reaction. — *Henry Spies*

Foundation Wall Cracks

Q. Within 60 days of pouring a concrete basement, some vertical cracks appeared. The cracks are between 1/32 inch and 1/16 inch wide, and extend from the top of the foundation wall down about 4 to 5 feet. My subcontractor tells me that the foundation is structurally sound, but the client is worried. Is this type of cracking normal, or can such cracks be avoided?

A. Many house foundations will exhibit small hairline cracking from curing and shrinking. If the cracks do not enlarge much more than their present size, the foundation should be fine. There are several possible causes of the cracks, including normal shrinkage from curing, and early backfilling of the foundation without bracing the walls. As concrete cures, it dehydrates and wants to shrink. Factors affecting whether cracks will appear on walls include the length of the walls, the number of wall penetrations, and the slump of the concrete.

Early backfilling can cause problems by placing extra pressure at the upper regions of the walls, where there is the least support. The pressure can cause small deflections in the upper wall areas, inducing cracks from the top of the wall down. You can check for wall deflection by running an offset string from wall end to wall end, verifying whether the distance from the string to the wall is consistent.

If the cracks continue to open up and lengthen as the walls cure, you should talk to your concrete contractor about your options for preventing water and insect penetration. If the cracks open up beyond 1/8 inch, the cracking may be caused by factors other than curing or early backfilling.

— *Jay Meunier*

Curing Concrete in Hot Weather

Q. I've recently moved my contracting business to the South. Concrete finishers here tell me that a slab poured outside on a hot day won't cure properly unless it's kept wet. They do this by having someone periodically spray the slab with water from a hose or they use a lawn sprinkler, once the concrete is set up enough that the top layer won't wash away. Is this really necessary? Would adding a set retarder to the mix accomplish the same thing?

A. To start with, a set retarder does nothing to help concrete cure. It simply makes the concrete set more slowly, but it does nothing to provide the water which concrete needs to cure.

For concrete to reach its full strength, it needs water to hydrate the cement. If it dries out, then the resulting concrete is soft, even chalky in an extreme case. This is most common on the surface of a slab. If it dries out even momentarily, it will be weakened (a condition called "dusting").

There are three important variables in determining how quickly the concrete will dry out: temperature, relative humidity, and wind speed. Thus, on a hot, dry, windy day, the concrete will dry quickly and that's when curing is most important. Most concrete has plenty of water when it is placed, so the key is to either prevent that water from evaporating or add enough supplemental water to make up for the evaporation. To prevent evaporation, you can use curing blankets, plastic sheeting, or membrane-forming spray-on curing compounds. Curing compounds can be reasonably effective when the evaporation rate is not too high. Using a pigmented compound allows you to see that you have complete coverage; using white pigment in hot weather helps by reflecting the sunlight.

A better way to ensure proper curing, though — as your local finishers have pointed out — is with water. The water can be ponded, sprayed, or misted onto the surface. To keep it wet, many concrete contractors lay down burlap to hold the water on the surface. But if you try this, don't let the burlap dry out or it could have a negative impact in hot weather by holding heat in. How long to keep the concrete wet depends on the air temperature and the mix: You want it to have reached sufficient strength on the surface before you let up. Typically, about seven days is sufficient with Type I cement; less is needed in warm weather. I always tell people that concrete is sort of like a baby with one important difference: When it is very young, if you keep it warm and wet (rather than dry) it will grow up to be a strong and responsible adult. Neglect it, and you'll have to live with a problem child for many years.   — *Bill Palmer*

Forming a Brick Shelf in a Concrete Foundation

Q. I'm planning to form a simple 4-foot stem wall foundation for a single-story garage using plywood, snap ties, and walers. It needs to have a brick ledge for the top 10 to 12 inches, which will show above grade. What's a simple, effective way to do this? Should I transition from a 10-inch wall, or can I use an 8-inch wall and reduce the thickness at the top? Would rigid foam board make a good form?

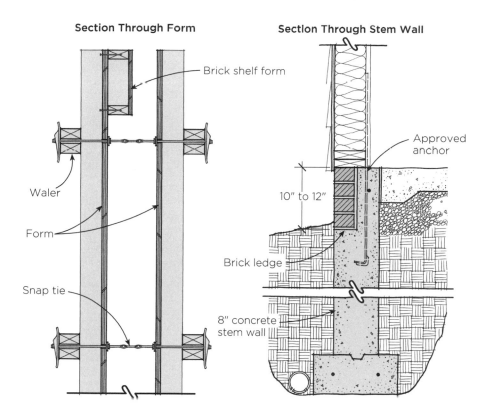

Section Through Form

Section Through Stem Wall

Brick shelf form

Waler

Form

Snap tie

Approved anchor

10" to 12"

Brick ledge

8" concrete stem wall

A. Sometimes rigid foam is used for forming pockets in complicated areas because it's so easy to shape. But it's fairly costly and has to be handled very carefully when you remove it from the formwork if you plan to use it a second time. A simple, inexpensive way to form your brick shelf is with 2x4s and 1/2-inch plywood (see illustration). Make a short "wall" with a top and bottom plate and uprights every 18 to 24 inches, and skin it with the plywood. It's easiest to build the shelf form separately, and then nail it into place as the forms are set.

Unless you have design or loading requirements necessitating a 6-inch stem in your wall, an 8-inch wall will work fine to carry a 4-inch brick shelf. The 4-inch stem makes it a little more cumbersome to place concrete in the wall, especially if it contains a mat of rebar, so plan on the pour taking a little longer than usual. When placing the concrete, consolidate the concrete underneath your brick shelf by using a concrete vibrator or rapping your formwork with a mallet. This will give the shelf area a clean, square edge and setting surface.

— *Jay Meunier*

Tying Into Concrete Block

Q. What is the best way to tie a new block addition into an existing block building?

A. Dowel the new footing to the original building with at least two #4 rebars. You'll need to dig to the bottom of the existing footing, drill it horizontally, and grout in rebars that extend about a foot into the new footing. The new footing should be at the same depth as the existing footing and on undisturbed soil. At the wall connection, vertically fasten a 3x3x^{1}/$_{4}$-inch angle to the original wall with expansion anchors (see illustration). Provide a bond breaker on each leg of the angle, and break out the end webs of the new blocks. Then, pack the cores with mortar around the angle, and seal the joint between the new blocks and the old wall with a good elastomeric caulk. This will provide lateral support to the wall but will allow some minor vertical movement without cracking. — *Henry Spies*

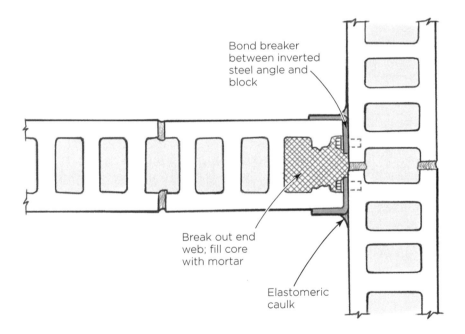

Bond breaker between inverted steel angle and block

Break out end web; fill core with mortar

Elastomeric caulk

Termite Defenses for Slab-on-Grade Foundations

Q. We've been told that it's a good idea to insulate the perimeter of slab-on-grade foundations, but we are concerned with termites in our area. What is the best way to detail the foundation to keep termites out?

A. Exterior foam on any foundation presents a real problem wherever termites are active. Termites tunnel right through the foam, making it nearly impossible to inspect the perimeter of the building for their passageways into the home. A 1/32-inch gap is all a termite needs to squeeze through to get inside a home, which makes it very difficult to detail any type of foundation to keep these insects out. Slabs are the hardest of all. While a monolithic slab is the best design alternative for a slab because it eliminates some of the gaps between the slab and a stem wall, termites can burrow beneath the slab and come up through the gaps around plumbing chases and electrical conduit.

The most common protection for slabs typically involves soil treatments beneath the slab and around the perimeter of the building. However, these treatments must be maintained on a regular basis, creating a long-term maintenance issue for the homeowner. Alternatives

Termimesh

For slab-on-grade foundations, Termimesh provides a barrier around plumbing penetrations. When concrete cures, it may shrink back from the pipe, allowing a tiny gap for termites to squeeze through. The stainless-steel mesh can be secured to plumbing with stainless-steel clamps (left) or installed beneath a foam block-out (right).

include borate treatments that target all the structural wood above the slab in order to rob termites of potential food supplies. These methods, which usually involve either buying pretreated framing lumber or spraying all the lumber prior to framing, have proved to be most effective against the Formosan termite, which may nest above ground.

The most promising termite protection available today is the Termimesh System (281/257-6558, www.termimesh.com), which is the only system available that actually blocks the entry of termites into the home. Developed in Australia and tested by U.S. Dept. of Agriculture Forest Service in Gulfport, Miss., for over a decade, Termimesh consists of a stainless-steel screen that is installed at the perimeter of the slab and at interior entry points through the slab.

At the slab's perimeter, Termimesh is bonded using a cementitious bonding agent that's painted over the mesh to seal the screen to the concrete. This will not completely prevent termites from entering a building through hidden gaps, since the insects can still build a passageway around the barrier. But like conventional metal shields that are installed correctly, the screen shield will force termites out into the open areas where their activity can be detected.

However, Termimesh offers a critical control that metal shields can't provide: It seals the tiny gaps around plumbing and conduit penetrations. The fine stainless-steel screen is sealed to pipes with stainless-steel clamps or laid beneath plumbing block-outs, and then embedded in concrete about halfway through the slab section (see photos, previous page).

Because termites primarily search for food by the scent of rotten or decaying wood, it's important to remove potential food sources from the job site and to protect wood on the house from moisture:
- Do not bury stumps and wood debris on site, and keep cutoffs and cardboard scrap out of the backfill.
- Remove wood concrete forms and stakes, and peel back the ends of Sonotube forms from the tops of poured piers.
- Control runoff with gutters and downspouts, backfill with well-draining material, provide good foundation drainage, and control site drainage. These practices will keep soil drier, robbing termites of the high soil moisture content they need for survival.
- Use only pressure-treated wood in contact with the ground.
- Be sure to hold siding and trim at least 8 inches above grade.

— *Clayton DeKorne*

Protecting Insulated Foundations From Termites

Q. We have termites in a timber-frame house whose slab-on-grade foundation is insulated with rigid foam (Illustration A). I understand that some exterminators will not guarantee treatments on homes with exterior foundation insulation. What is the best way to detail a foundation to help keep termites out?

A. Pest-control professionals are right not to make promises when faced with an insulated foundation, because it's impossible to inspect inside and behind the insulation.

A 1/32-inch gap is all termites need to sneak into a house. To eliminate these gaps, begin by choosing a foundation type and materials that present the fewest possible entryways.

Of all foundation types, slabs-on-grade are the most vulnerable. In your design, for example, termites can slip through the foundation/slab joint, inside the hollow concrete blocks, or between the insulation and foundation (Illustration A). Monolithic slabs, where grade beam and slab are cast in one pour, lack joints and therefore present fewer entry routes than slabs supported on foundation walls. Likewise, a cast concrete stem wall, with its solid center and reinforcement to minimize cracking, is better than a hollow block wall. However, block walls can be made more termite-resistant by capping them with reinforced concrete or solid blocks, or by plugging the hollows in the top course with mortar.

Making any foundation termite-resistant requires treating the soil under the slab and surrounding the foundation with a termiticide.

Treating the soil under the slab, of course, is best done before the slab is cast, and treating the surrounding soil should be done during finish grading.

Termites prefer wetter soils, so make sure water is directed away from the foundation by properly sloping finish grades, and by using gutters and downspouts. Also, keep the below-slab drainage pad higher than the outside soil by elevating the slab surface at least 8 inches above the finish grade. Reinforce the slab to minimize cracking, and design utilities so that the slab penetrations are minimized or eliminated. All penetrations and joints should be sealed with roofing-grade coal-tar pitch. Also, make sure that all wood (read "termite food"), such as stumps, grade

A. Slab-on-Grade With Stem Wall

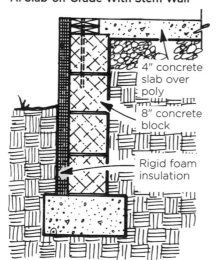

4" concrete slab over poly

8" concrete block

Rigid foam insulation

This slab-on-grade foundation is a poor choice in areas where termites are a problem. Termites can slip through the foundation/slab joint, inside the hollow concrete blocks, or between the insulation and foundation.

B. Monolithic Slab-on-Grade

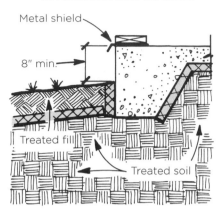

C. Basement or Crawlspace

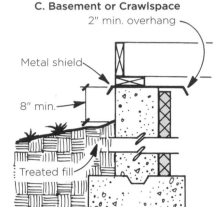

A monolithic slab (left) or a cast concrete wall (right) presents fewer entry routes than a slab resting on a stem wall. It's important to leave at least 8 inches of the foundation exposed above grade. Although some insulation value is lost, this allows an exterminator to inspect for the presence of termites.

stakes, formwork, and scraps, is removed and disposed of off site, and not buried during backfilling.

When detailing the insulation, you may have to sacrifice some energy efficiency in return for being able to properly inspect for termites. The details in Illustrations B and C show two ways to leave the foundation exposed for inspection and also reduce the number of possible entries. The metal shields shown in these details should be thought of as only one part of a home's anti-termite defenses. They're seldom fabricated and installed as tightly and as carefully as is really necessary. — *Stephen Smulski*

Do Termite Shields Work?

Q. Will termite shields actually deter the insects from entering a house?

A. Termite shields are breeders of false confidence. Termite shields do not prevent the entry of termites, but supposedly force the termites to build a tube around the shield, so their presence can be detected during an annual inspection. According to conventional design, an effective termite shield is a piece of metal flashing that projects at least 2 inches below the woodwork on each side of the foundation, with the outer edge bent down at a 45-degree angle. All joints and openings, such as around anchor bolts, must be sealed. But because of porch and patio slabs, interior finished walls in basements, and other house details that interrupt the barrier, I have never found a house fully protected by termite shields. Soil poisoning, baiting, and new stainless-steel mesh systems are the only effective protection against subterranean termites that I know of. *— Henry Spies*

Drywood Termites

Q. I work in and around Albuquerque, N.M. Do drywood termites cause trouble in this area? If so, what measures can one take to prevent damage?

A. Drywood termites are present in New Mexico, and they can definitely cause trouble. They are found in a very narrow region along the southern fringe of the U.S., from California to Florida. Drywood termites do not live in the ground like the more well-known subterranean termites, and they don't multiply as fast.

In the U.S., drywood termites cause far less damage than subterranean termites, but their ability to live in dry wood, without outside moisture or contact with the ground, makes them particularly troublesome. The destruction caused by drywood termites does not proceed rapidly, but over the course of many years they can completely destroy the timbers in a home.

Drywood termites are seldom seen and are difficult to detect. Signs of the presence of drywood termites include tiny termite fecal pellets and "kickout holes" in the wood, which are the size of a BB. Drywood termites can be transplanted from one building to another in boxes, furniture, lumber, and other infested wooden objects. It is important to inspect lumber carefully before you build with it. Remove wood scraps, debris, old brush, and stumps from your building site. Keep exposed wood painted, since paint is a fairly good barrier to infestation by drywood termites.

Even better, use treated wood where possible. Pesticides can be used to get rid of existing drywood termites in a home. You can either spray the nests directly or fumigate the house, which is both very effective and very dangerous. However, reinfestation after fumigation is possible, since fumigation does not leave any residue behind. Spraying or fumigation should be left to a licensed exterminator. Call your state pest control office for a listing of licensed professionals. *— Paul Fisette*

Framing

2

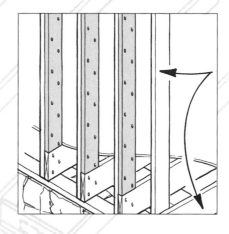

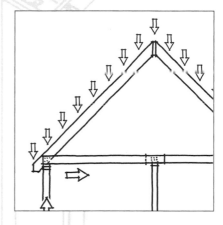

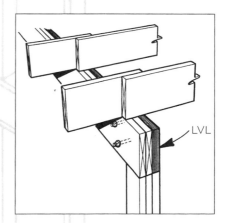

Squash Blocks and Web Stiffeners

Q. I've been using wood I-joists for my floor framing. The designs have called for 1x4 crush blocking inserted between the top and the bottom 2x2 flanges and nailed directly to the OSB web. I've also heard of using 2x4s nailed to the face of the 2x2s, instead of being inserted between them. Which way is better?

Squash Blocks

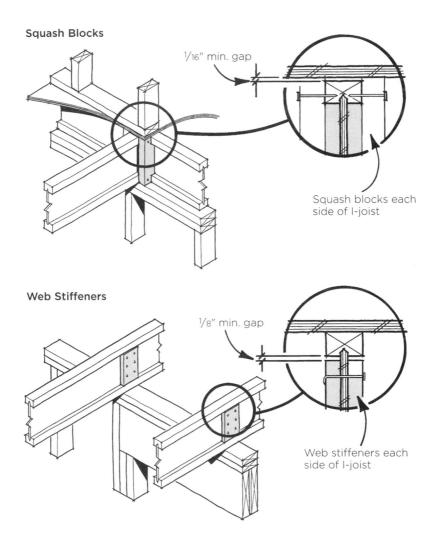

1/16" min. gap

Squash blocks each side of I-joist

Web Stiffeners

1/8" min. gap

Web stiffeners each side of I-joist

A. It sounds like you're confusing squash blocks with web stiffeners. Squash blocks are blocks of wood that are slightly taller than the depth of the joist (see illustration). The function of squash blocks is to transfer loads created by walls and roofs above to a sill or wall plate below the I-joists. Because these loads in effect bypass the I-joist, the squash blocks also prevent "knifing" of the thin web into the top and bottom flanges.

Web stiffeners, on the other hand, are blocks of plywood or OSB that fit between the flanges of an I-joist. These blocks are used at points of support and are fastened with nails right up against the I-joist web. However, web stiffeners are not fit tightly between the top and bottom flange. Typically, a space is left between the top of the stiffener and the underside of the top flange. Web stiffeners reinforce the I-joist and prevent it from buckling. Sometimes they are required as fillers so that there is solid nailing for joist hangers. — *Paul Fisette*

Squeakless Floors

Q. How can I avoid floor squeaks with a wood joist system?

A. The best way to avoid floor squeaks is to follow these steps:
- Use dry lumber for the joists.
- Use a tongue-and-groove subfloor.
- Use a mastic-type construction adhesive, such as PL-200, between the joists and the subfloor, and on the tongued edges of the subfloor.
- Nail any underlayment tightly with ring-shank nails, but space the underlayment nails so they do not hit the joists.   — *Henry Spies*

Best Subfloor Adhesive for Wet Conditions

Q. Which construction adhesives are best for gluing plywood subfloors to joists under wet conditions?

A. All construction adhesives that are labeled AFG-01 or ASTM D 3498 are capable of doing the job. One of the tests to pass for these specifications is a wet lumber test. Solvent-based construction adhesives typically dissipate the water better than latex- and urethane-based adhesives.
— *Mark A. Roberts*

Notches and Holes in Joists

Q. What's the largest hole I can drill in a 2x8 joist? What about notching? Also, how close to a bearing point can I drill a hole?

A. The rule that most codes use is that holes can't be any closer than 2 inches from the top or bottom of the joist, and cannot be within 2 inches of any other hole. If the joist is notched, you can't drill the hole within 2 inches of the notch. The hole's diameter can't exceed one-third of the joist depth. So in your case, the largest hole would be 2³/8 inches.

Notches in the top or bottom of a joist can't exceed one-sixth the depth of the joist and can't be longer than one-third the joist depth. Notches must not be made in the middle third of the span. If a notch is made at the very end of the joist, it can't exceed one-quarter the joist depth. As always, check your local code to make sure it's in keeping with the norm.
— *Paul Fisette*

Joist Size	Maximum Hole	Maximum Notch Depth	Maximum End Notch
2x4	None	None	None
2x6	1¹/2"	7/8"	1³/8"
2x8	2³/8"	1¹/4"	1⁷/8"
2x10	3"	1¹/2"	2³/8"
2x12	3³/4"	1⁷/8"	2⁷/8"

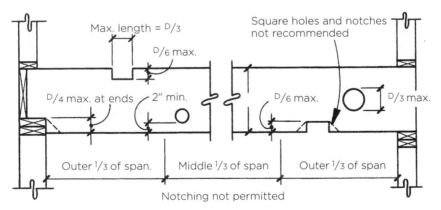

Do not notch a span's middle third where bending forces are greatest. For all calculations, use actual, not nominal, dimensions.

Drilling Holes in Wood I-Joists

Q. I need to run wires through Boise Cascade I-joists. The joists come with prepunched holes, but the holes don't line up. I'd like to drill some 1-inch holes in the joist webs. Are there any limitations on where these holes can be drilled?

A. A hole up to 1 1/2 inches in diameter can be drilled anywhere in the web of a BCI joist, as long as it is located at a distance equal to at least two diameters away from any other hole. In other words, a 1-inch hole needs to be at least 2 inches away from any other hole. This is generally true of most I-joists, but check with manufacturers for specific recommendations.

— *Kevin Pelletier*

Floor Stiffness: I-Beams vs. Solid Lumber

Q. Is a wood I-beam floor bouncier than one built with dimensional lumber?

A. Wood I-beams are actually stiffer than dimensional lumber of a comparable size. But you have to distinguish between "bounce" and "vibration." Because wood I-beams are lighter, they tend to vibrate more easily, giving the floor a hollow feel. This generally won't be a problem if a drywall ceiling is installed below the floor. If the floor is open below, install strapping to the bottom of the joists on 6-foot centers to reduce the vibration.

To reduce bounce, select wood I-beams for an allowable deflection of 1:360, not 1:240. If you are building for a greater anticipated load than normal (a water bed, a grand piano), go to the next deeper joist than required by the design tables.

— *Henry Spies*

Sizing Steel Girders

Q. I've read that you should limit deflection of built-up wood girders to L/600 so that the floor joists won't vibrate excessively. My question: Does this also apply to steel girders, or is it mainly a wood-related issue, owing to the inherently greater flexibility of wood beams? In other words, would a steel beam sized for L/360 deflection be less bouncy than a wood beam sized for, say, L/480? Or should I absolutely make sure that all my steel beams are sized for the L/600 limit?

A. The forces that induce floor vibration and the reaction to it are blind to the beam or girder material. Each girder material, whether wood or steel, has an elasticity, E, and that design value is used to predict deflection. The designer sets limits on live-load deflection. A wood beam and a steel beam both designed to exactly meet the L/360 deflection limit should perform the same, assuming the weight of the girders themselves has been accounted for and that both are fastened securely to their support points.

Considering the price of steel (and the cost to repair a bouncy floor), you should upgrade all girders to meet a live-load deflection limit of L/600 if vibration is a sensitive issue for your clients.

— *Frank Woeste and Dan Dolan*

Reducing Floor Vibration

Q. In the search for a bounce-free floor, does either solid blocking or cross-bridging do anything to reduce the vibration caused by foot traffic? I haven't been able to sense a difference between blocked or bridged floors and those without blocking or bridging. Consequently, I haven't used either technique on my floor systems for several years.

A. We actually evaluated the effect of bridging or blocking on the vibrational performance of solid-sawn joists and I-joists in the laboratory. Full-scale floors were tested with and without bridging or blocking installed. Based on key indicators of floor performance, bridging and blocking had only a minimal effect on reducing annoying vibration. Some general guidelines for controlling bounce include the following:

Shorten the span. In general, shorter spans make for stiffer floors. For example, if the L/360 span table tells you a joist of a given size, grade, and species will just barely work for your span, then shorten the span by adding a girder near the center of the original span. The resulting floor will vibrate less.

Increase the joist depth one size. If the code requires a 2x8 at 16 inches o.c., then use a 2x10 of the same grade and species. Or use a 14-inch-deep floor truss when a 12-inch-deep truss would meet code requirements. This may not be the most cost-effective solution in every case, but it's easy to remember and will save time and worry.

Reduce joist spacing. Probably the least efficient way to improve floor performance is to reduce the on-center spacing — 16 inches to 12 inches, for instance. Occupants feel "bounce" as a result of a foot impacting an individual joist. But even at 12 inches o.c., the joists are not close enough for the shock of a foot to be carried by two joists.

Glue and screw the sheathing. Floor sheathing should always be glued down. Screws work better than nails for long-term bounce control.

— *Frank Woeste and Dan Dolan*

A Fix for Bouncy Floors

Q. As part of a remodel, I have to replace the carpet on a floor framed with 2x10s on 12-inch centers. The joist span is 20 feet, and the floor bounces so badly that things on the table shake when someone walks through the room. Is there a good way to fix this problem before recarpeting?

A. A possible solution, which I have used successfully in several cases, is to add a layer of 3/4-inch or thicker OSB over the entire floor, glued and screwed 6 inches on-center in both directions. Make sure the screws are long enough to fully penetrate the new OSB and the underlying subfloor. When possible, using longer screws to penetrate the joists will help some, but the important thing is to make sure the new subfloor is well "clamped" to the existing. Offset the joints between layers.

This technique works for one of two reasons, depending on the individual floor. One is that the extra OSB increases the stiffness of the floor, making it vibrate at a higher frequency that is not perceived as annoying by most people. In some cases, however, the increased mass of the OSB actually lowers the frequency of vibration, but it doesn't tend to matter because the increase in mass also increases the inertia of the floor. In other words, the heavier floor is harder to start vibrating, so the problem goes away.

One problem with this approach, however, is that you gain floor height, which may not be acceptable. — *Dan Dolan*

2x4 vs. 2x6 Walls

Q. Is there any structural difference between a wall framed with 2x4s 16 inches on-center and a wall with 2x6s 24 inches on-center?

A. There is almost no difference in the bearing capacity (the wall's ability to support a compressive load), which is how most walls are loaded. Bearing capacity is a function of the footprint area of all the studs in a wall. For example, a 4-foot section of wall would have three 2x4s, but only two 2x6s. The total bearing area of three 2x4s is $15^3/4$ square inches; two 2x6s have a bearing area of 16 square inches.

In bending, however, such as from a wind load, a 2x6 wall is considerably stronger.

In tall walls, where column buckling might be a factor, a 2x6 wall would be stronger if structural sheathing was used. Structural sheathing provides lateral support to the $1^1/2$-inch dimension of either 2x4s or 2x6s, but the greater width of the 2x6 makes it stiffer in that direction. If the sheathing is not structural, both wall systems would have equal resistance to buckling, since the buckling would occur across the thickness of the member. *— Henry Spies*

Pressure-Treated Plates on Slabs

Q. Do you need to use pressure-treated lumber for the bottom plates on interior partitions over a slab?

A. Code requires pressure-treated lumber wherever framing is in contact with foundation concrete. Another very good reason to have pressure-treated lumber plates is to deter termites. Termites can enter through a crack as small as $1/32$ inch, and the pressure-treated plate can discourage their entry. *— Henry Spies*

Which Way With the OSB?

Q. I recently inspected a home that had $3/8$-inch OSB corner bracing. The surface stamp indicated that the strength axis of the panel ran in the long direction (see photo). I assume this means the panel should be applied with the long dimension across the studs, yet the builder had installed the long dimension parallel to studs. I called APA to check and was told the stamp only applies for roof sheathing and not to corner bracing. Why is the strength axis important for roof sheathing but not for corner bracing?

A. There are two properties of OSB panels at play here: its strength in shear and its bending strength. When using OSB for corner bracing, shear strength — the ability of the panel to resist lateral racking movement — is the issue, and the panel's shear strength is not affected by its orientation. But place that same sheet up on the roof across rafters 24 inches on-center and stand in the middle of the span between two rafters: now the bending strength of the panel is at play. That's what the arrows refer to. If you look carefully at the sheet of OSB in question, you'll see that most of the fibers are oriented parallel with the length of the sheet, making that the strongest axis in bending. — *Scott McVicker*

Shear Strength of Gypboard

Q. I've heard that engineers give no structural "credit" to gypboard, but I know it greatly stiffens partitions when I nail it up. How much shear strength does drywall really have, and why not credit it in the design?

A. As you suspect, properly fastened gypboard does have the capacity to resist racking and/or lateral forces. Table 25-1 of the 1997 Uniform Building Code gives shear values for both gypsum wallboard and gypsum sheathing. In fact, the allowable lateral force on a wall with fully blocked 5/8-inch gypboard on both sides nailed at 4-inch centers (350 plf) actually exceeds that of a wall with 1/2-inch Structural I plywood fastened with 10-penny nails at 6-inch centers (340 plf). Be careful, though: if you are working in seismic Zones 3 or 4, note that even with fully blocked edges you must reduce the allowable lateral load on gypboard by 50%.

As to crediting the design for the strength of the gypboard, this decision is based on the materials selected for the particular structure. If you build a house with rigid-foam insulation panels on the exterior (under finishes) and gypboard on the interior, then the gypboard is the lateral force-resisting material. However, if the interior gypboard is combined with plywood sheathing on the exterior (or with diagonally braced structural steel studs), then the strength of the gypboard is discounted. In the latter case, the plywood is considered the primary lateral-force-resisting material because of its greater strength and stiffness. In both instances, the designer must make certain that the primary lateral-force-resisting material is sufficiently fastened to the framing to resist the total lateral load despite the presence of other secondary materials.

In reality, it is the combination of all the primary and secondary materials that will resist the applied lateral loads. However, should the loading persist, the repetitive cycles of load/release will cause fatigue of the weaker materials (like gypboard) until essentially only the primary lateral material remains functional. If we were to credit the strength of the gypboard toward the total lateral load (and reduce the plywood nailing accordingly), our structure would lack critical capacity after the time when the gypboard had yielded. This is the reason gypboard receives no credit for its strength.   — *Scott McVicker*

Framing With L-Corners

Q. The local framing inspector told us that our L corners are not acceptable. He says that an L-corner is not as strong as a corner framed with two studs separated by blocking. Is he right?

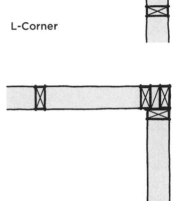

L-Corner

Built-up Corner

A. It's true that the illustrations for corner framing on braced wall panels in the International Residential Code show a built-up corner. But that may only be a historical convention. As long as you install nailed structural sheathing, an L-corner (also called a "California corner") is certainly strong enough. Extra studs at the corner of a house are unnecessary for supporting the building's vertical load and provide little significant benefit to improve a frame's racking resistance, which is provided by the nailed sheathing.

When it comes to supporting the vertical (compression) load from upper stories and the roof, the stud at the end of a wall carries about half the load of most other studs. A stud in the center of a wall carries 16 inches of load (8 inches on either side of the stud), while a stud at the end of a wall carries only 8 inches of load — from the stud to a point halfway to the next stud in the wall. Furthermore, most gable ends have studs that redistribute some of the roof loading at the end of the wall.

In most cases, the easiest way to resolve a dispute with an inspector is to agree with the inspector. But in this case it's worth trying to persuade the inspector, since there is an energy conservation benefit to framing with L-corners.
— *Paul Fisette*

Fastening Studs With Screws

Q. On a remodel job, I used screws to fasten studs and blocking in a nonbearing partition. An inspector told me that only nails, not screws, could be used in this application. Is this correct?

A. The model building codes accept nails as structural fasteners and provide the required nailing schedules for each type of nail. When something other than nails will be used, a builder must look at the code section on "alternate" fasteners. That section of the code covers fasteners like staples, pneumatic nails, and screws.

The problem with most screws is that they are hardened and brittle. Some screw manufacturers have had their fasteners tested under the National Evaluation System (NES) or by some other code-recognized evaluation system. The evaluation reports provide performance data and sometimes equivalent strength values.

The acceptance of screws for framing, even in a nonbearing partition, is up to the discretion of your local building official. In most cases, the official is likely to demand that any screws used must prove their performance with an evaluation report, available from the manufacturer of the screw. — *Paul Fisette*

Shoring a Balloon Frame

Q. I am rehabbing an old two-story balloon-framed house with a crumbling limestone foundation. I need to jack the walls slightly to repair the foundation. The floor joists rest on a 2-by leveling plate, and the 20-foot-tall studs are face-nailed to the sides of the joist ends (see illustration). With platform framing, I would ordinarily run a temporary beam under the joists as close to the foundation wall as possible and place my jacks under the beam. If I do that with this frame, I am concerned that the weight of the walls, which are carrying the second floor and roof, will shear the old face nails. Any suggestions?

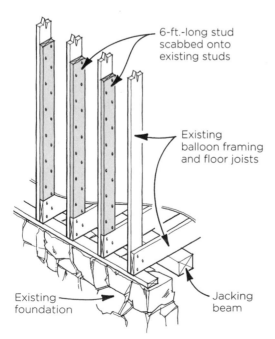

6-ft.-long stud scabbed onto existing studs

Existing balloon framing and floor joists

Existing foundation

Jacking beam

A. Here's a simple technique that will work if you can get into the wall cavity. Scab on 6-foot lengths of stud material along the bottom of each wall stud (for a single-story house, I might go with every other stud). Make sure the bottom of the sister member sits squarely on top of the joist, and nail it off with ten or twelve 16-penny nails. (Use duplex nails if you plan to remove the scabs. In my climate, uninsulated wall cavities are common, so I just leave the 6-foot members in place permanently.) Now you can jack in the manner you described. — *Mike Shannahan*

Rusting Steel Studs

Q. We are currently building a steel-framed house near the ocean in California. Our site is very close to the water, and I am concerned about rust problems developing in the steel framing. We have already noticed small surface rust spots on some of the steel tracks on the floors. It seems as though these rust spots are being caused by the small particles of metal shavings left in the tracks either from drilling through the studs or where they were screwed together during assembly. I have also noticed that rust is forming on the ends of cut studs. Will the rust stop after the house is closed up, or will the rust spots grow over the years?

A. All structural steel studs should be galvanized. When you build in a coastal environment, you should specify a grade called G-90. G-90 steel has a heavier zinc coating to provide extra protection in hostile environments.

Rust on a steel stud can be compared to rot on a wood stud. When a wall is properly constructed, there is little likelihood of a stud rusting to the point of failure. But it is important to use proper wall construction that shields the framing from exterior elements and minimizes the chance of condensation within the wall cavity.

Normal cutting and drilling of steel studs removes the zinc coating in the cut area. However, in most cases the coating adjacent to the cut will "sacrifice" itself to protect the cut area. If you are concerned about areas of extensive rust, such areas can be cleaned with a wire brush and then sprayed or brushed with a zinc paint called ZRC (ZRC Worldwide, 800/831-3275; www.zrcworldwide.com). The zinc in ZRC is the same material used to galvanize the studs. ZRC can be purchased in a good paint or hardware store. — *Paul Fisette*

Rafter Framing With Unequal Wall Heights

Q. How do you figure rafter lengths for a gable roof when one wall is 10 feet high and the other is 8 feet high (Illustration A)? The span of the building is 24 feet. We want the roof pitch to be the same on both sides.

A. Using a sample roof pitch of 6:12, the first step is to calculate the rafter run for the 2-foot difference in rafter heights. This is done by dividing the difference in inches (24) by the rise ratio of 6:12 (.5), to arrive at a 48-inch run (24 ÷ .5 = 48). Next, add 48 inches to the 24-foot span to create a phantom 8-foot-high wall outside the 10-foot wall, as shown in Illustration B. Now, figure the rafter lengths for the 8-foot wall using the 28-foot span. The rafters on the 10-foot-high side can be most easily figured by "installing" a 10-foot-high phantom wall 48 inches in from the real 8-foot-high wall (Illustration C), and calculating the rafter lengths for a 20-foot span. — *Will Holladay*

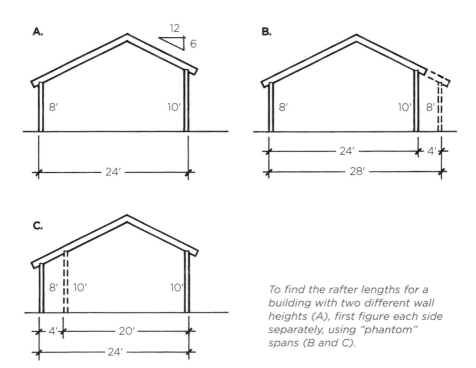

To find the rafter lengths for a building with two different wall heights (A), first figure each side separately, using "phantom" spans (B and C).

Staggered Rafters at Ridge

Q. I am framing a gable roof with common rafters attached to a 2x12 nonstructural ridge board. Must the rafters on either side of the ridge line up, or can they be staggered along the ridge board?

A. Rafters can be safely staggered across a ridge board in most situations. Section 802.3 of the IRC "Framing Details" does not explicitly require alignment of rafters on opposite sides of a ridge board. There are certain advantages to aligning the rafters, however, and unless you have a compelling reason not to, I recommend that you line them up. Not only does it make framing ceilings and collar ties easier, but the roof framing simply looks more professional when the rafters align. Having said that, in many cases, staggering the rafters does not create excessive stresses in the ridge board.

To satisfy myself that the ridge board will not be overstressed, I analyzed a somewhat conservative scenario. I chose a gable roof at a 6:12 pitch with 2x10 rafters spanning 16 feet from exterior walls to a 2x12 ridge board, to carry design loads of 30 pounds per square foot (psf) live load and 15 psf dead load. In this scenario, the unbraced rafter spans are relatively long, the design loads relatively high, and the roof pitch relatively shallow. I assumed that the ridge board is #3 southern yellow pine, the weakest grade available. Finally, I framed my sample roof with the maximum possible offset; each rafter is 8 inches from the two nearest rafters on the opposite side of the ridge. In spite of all of these conservative assumptions, I found that the 2x12 ridge is not overstressed in shear or in bending, and that predicted deflections are negligible.

Whenever you must splice the ridge board, line up rafters from both sides centered on the splice so that 3/4 inch of rafter material pushes on the ridge on each side of the break. As a rule of thumb, don't offset rafters under any of the following circumstances:
- when rafter spacing exceeds 16 inches
- when the roof live load exceeds 30 psf
- when the roof pitch is 4:12 or lower — *Christopher DeBlois*

Which Side of OSB Goes Up?

Q. When OSB is installed on roofs, which side should face up — the slick side or the rough side?

A. The rough side is installed facing up, in order to provide a safer walking surface for workers during construction. — *Paul Fisette*

When to Double Rafters at Openings

Q. When you remove one rafter to install a skylight, do you have to double the two rafters at the sides of the opening? What about the headers?

A. When you're framing a skylight that requires you to cut no more than two rafters, a common rule of thumb says to double the rafters on either side of the opening and use double headers. This is often required by building inspectors, but in fact, it's a very conservative guideline that often results in unnecessary framing. Only in rare cases are the doubled headers required. And in many cases, particularly when 2x10 or 2x12 rafters are chosen for insulation thickness rather than for strength, doubled rafters may not be required. In my experience, the inspectors also often insist on joist hangers at the headers, but these are also rarely needed. Usually, an adequate number of 16d nails (as many as eight nails for a 2x12 connection, depending on the loads) can handle the reaction forces. Watch out, though: as the opening size increases to the point where you're removing three or more rafters, even doubling the perimeter framing may not be sufficient.

— *Robert Randall*

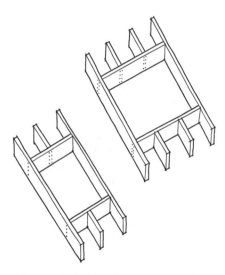

If deep rafters have been selected to accommodate insulation requirements rather than structural requirements, it may be unnecessary to double trimmer rafters. It would rarely be necessary to double the opening headers.

Using Purlins and Struts to Reduce Roof Span

Q. Does using purlins and struts at midspan allow you to cut the roof span in half compared with what's given in a rafter table?

A. Yes. Actually, in theory, a rafter with purlin support at midspan could be a little longer than twice the maximum allowable single span length. This is due to the effect of moment continuity across the support. This means that the roof load on one side of the purlin has a slight lifting

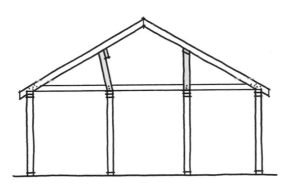

effect on the other side of the purlin.

When using purlins, you must be careful that the struts are properly supported by a bearing wall, beam, or other suitable support. Always carefully trace the load path down to the ground, verifying the adequacy of each element. To be most effective, struts should be installed as close to vertical as possible so as not to create lateral forces that have to be dealt with. This will depend, of course, on the location of the bearing wall below. And keep in mind that when the struts get longer than 6 feet, they may require lateral bracing. — *Robert Randall*

Purlins and struts reduce rafter spans just as bearing walls do. The struts (left) or a kneewall (right) should be installed as close to vertical as possible, and must be properly supported below.

Strength of Header Hangers

Q. How do header hangers (see illustration) compare, from a strength standpoint, with traditional supporting jack studs?

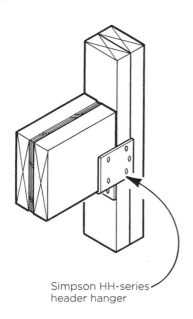

Simpson HH-series header hanger

A. If the header and jack stud are perfectly cut so that the header bears completely on the jack, the jack stud would have about three times the capacity of a header hanger in download. It should be noted, however, that the strength of the header hanger may still be sufficient for a particular installation. The advantage of header hangers comes in eliminating the jack stud altogether, thereby allowing better insulation properties, saving on lumber, and allowing slightly wider openings in tight framing situations. Simpson's HH series hangers, with their $2^{1}/8$-inch seat, will also provide support for a header that has been cut slightly short. — *Robert Bouchet*

When Are Collar Ties Needed?

Q. Collar ties don't seem necessary in attics where the rafters come all the way down to the ceiling joists. Can you remove some of them to create headroom?

A. The most common reason for installing collar ties is to prevent rafters from spreading apart under load. However, in a conventionally framed peaked roof, like the kind you describe, collar ties would probably serve little or no function, since the attic floor joists serve as ties to prevent the rafters from spreading. Note that the connections between the rafters and the joists must be adequate, and that the overlapping joists at midspan must also be properly nailed.

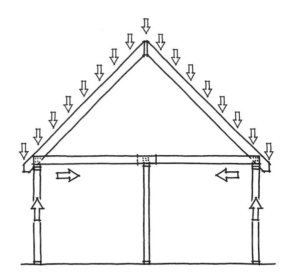

Collar ties are usually not needed in conventional gable roof attics, as long as the floor/ceiling joists are properly connected to handle the tension forces created by the outward thrust of the roof.

There are some exceptions, however, when collar ties might be useful even in a conventional attic roof. For example, very long rafters in a relatively steeply pitched roof (slopes above 6:12, for instance) may benefit from a stabilizing effect if adequately connected collar ties are installed on every rafter pair. In this case, the collars serve not as ties but as spreaders.

Also, in high-wind situations with lower pitched roofs, collar ties may help hold the ridge assembly together, although steel strap ties installed just below the ridge board would probably work better.

My call is that in the vast majority of such cases, collar ties can be removed with no detrimental effect. In most of the cases I have observed, the existing connections between the collar ties and the rafters are inadequate to provide any meaningful beneficial effect anyway. — *Robert Randall*

Retrofitting Plywood Clips

Q. A friend of mine recently tried to sell a home. The buyers' inspector discovered that no plywood clips had been installed on the roof sheathing. The roof has trusses 24 inches on-center and either 1/2-inch plywood or 7/16-inch OSB sheathing. Does anyone know a quick and easy fix for this?

A. I don't know of a "quick and easy" fix for this, but here is a method that I have used. Install 2x4 blocking across each sheathing joint from rafter to rafter. These blocks can be nailed through the rafter or truss or toe-nailed in place. Apply construction adhesive to the blocks to glue them to the decking. Though tedious, this simple repair works well. However, it is difficult to do on the first row of decking on low-slope roofs and impossible in some places due to framing members, cathedral ceilings, and so forth. *— Robert Allison*

Nailing Sheathing on Valley Roof Trusses

Q. We're framing a gable-roofed house with roof trusses. The roof includes a bump-out with a smaller gable roof that intersects the main roof, forming two valleys. A valley set will be installed on top of the main roof sheathing. The sheathing along the valley doesn't have blocking for continuous nailing, and I'd rather not nail sheathing to sheathing. Should I install blocking between the framing members of the valley set to provide better nailing?

A. Installing blocking is not necessary, since the sheathing in a valley doesn't need to be nailed between framing members. The same situation occurs at a ridge; although a truss roof, unlike a conventionally framed roof, has no ridge board, no blocking is necessary at the ridge. Contractors who feel more comfortable with blocking can always install it, but it is unnecessary. *— Don Richardson*

Roof Truss Spacing

Q. We see ourselves as quality builders. We frame everything 16 inches on-center, including roof trusses.

Most visitors to our job sites feel that this is overkill and that we're wasting the customer's money. Is putting trusses 16 inches on-center with $5/8$-inch sheathing and $1/2$-inch drywall on the ceiling a thing of the past?

A. I am not aware of any performance issues regarding $1/2$-inch drywall installed on trusses at 16-inch versus 24-inch centers. If interior moisture were improperly managed — for example, if a clothes dryer were vented into a finished garage area instead of to the outside, a 16-inch on-center truss spacing would be more forgiving of the bad practice.

But in general, it doesn't matter if the truss spacing is at 12, 16, or 24 inches. What dictates the truss design is the roof design snow load — typically 20 or 30 psf in most of the U.S. Once the builder or architect specifies the loads (for example, 20-10-0-10, for top chord live, top chord dead, bottom chord live, and bottom chord dead loads), the truss spacing, and the desired shape of the truss, then truss engineering takes over.

The engineering design — usually computer generated — dictates the size and grade of lumber required in the chords and webs. There are 10 grades of 2x4 southern pine, for example. If the builder requests a 16-inch on-center spacing, a lower grade will be used that meets the structural requirements. Conversely, builders who request the 24-inch on-center design should get a higher grade of lumber relative to the 16-inch on-center design. Higher grades of lumber obviously cost more than lower grades, so it becomes a trade-off between spacing and grade. In general, the wider 24-inch spacing is the most economical for residential construction. — *Frank Woeste*

Truss Uplift Solutions

Q. What would cause the sole plate of a nonbearing partition wall to pull away from the subfloor? The wall is nailed into roof trusses where they cross it above. There is no sag in the floor system below.

A. If your floor is framed with green lumber, shrinkage of the floor joists could contribute to the problem you describe. But the primary cause is probably truss uplift. Lower truss chords that touch the warm ceiling dry out and shrink during the heating season, while the cold upper chords gain moisture and expand. The whole truss then curves like a bow, rising at the center. If the truss is nailed to the wall, it pulls the wall with it. The moving truss can also create drywall finish cracks in the corners where walls and ceilings meet (see illustration, next page).

Whether floor joist shrinkage or truss uplift is causing the problem,

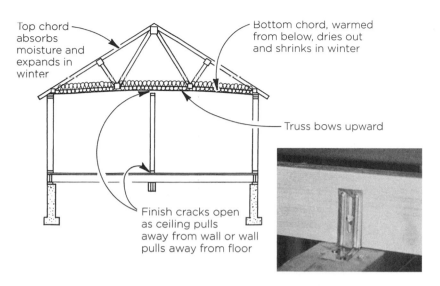

Top chord absorbs moisture and expands in winter

Bottom chord, warmed from below, dries out and shrinks in winter

Truss bows upward

Finish cracks open as ceiling pulls away from wall or wall pulls away from floor

Shrinkage in bottom chords and expansion in top chords cause trusses to flex upward at the center, lifting walls and ceilings with them. Prevent the problem with truss clips (shown in photo), which allow the truss to move up and down with seasonal moisture changes, while restraining movement in any other direction.

the way to prevent it is the same: Instead of nailing trusses to wall plates, use truss clips as shown in the photo (available from Simpson Strong-Tie, 800/999-5099; www.simpsonstrongtie.com). The clips allow each truss to flex up and down freely while preventing it from moving side to side. Also, don't screw the ceiling drywall to the truss where it passes over the partition. Instead, hold your screws 12 to 16 inches back from the partition and screw the edge of the ceiling drywall to 2-by nailer blocks that have been fastened to the wall top plate between the trusses. This gives the drywall enough room to flex at the joint and prevents a crack from forming.

We usually see these movement problems in the first couple of heating seasons. After that, the houses seem to settle down and the problems lessen. In any case, the wall should drop back down in the spring, so wait until then to fix the problem. Whatever you do, don't shim under the wall during winter when the crack appears — if you do, when the roof settles back down in spring, you'll have created a bearing situation in the center of the truss where there isn't supposed to be one.

Another option, useful in retrofits, is to install cove or crown molding at the corner. Make sure to attach the molding through the ceiling to the trusses but not to the partition. — *Dave Utterback*

Crown Detail Hides Truss Movement

Q. We all know truss uplift can be a problem at the intersection of ceilings and interior walls. Is there a way to install crown molding so that you don't see a paint-line reveal when the ceiling moves?

A. Truss framing can lead to major separations at the intersection of ceilings and interior walls. If your crown molding is assembled from at least two components that can slide independently, it should help disguise some truss uplift movement, rather than be part of the problem. The aim is not to resist the movement but to hide any gaps as much as possible. Where truss uplift is a major problem — that is, where a truss-framed ceiling moves seasonally an inch or more — the solution suggested here may need to be scaled up.

I typically install crown molding with a rough base underneath it. This nailing base is triangular in section, with the outer face fitting between the back of the crown molding and the wall. I also usually install at least one additional piece of flat trim stock — which can be square-

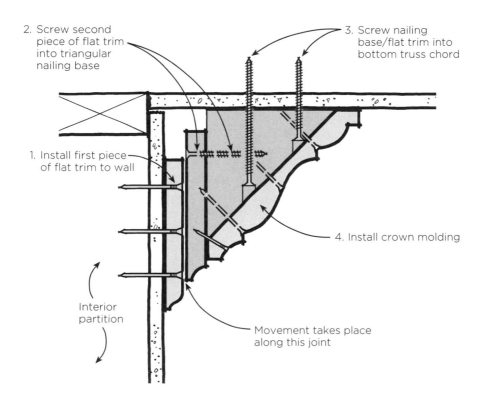

2. Screw second piece of flat trim into triangular nailing base

3. Screw nailing base/flat trim into bottom truss chord

1. Install first piece of flat trim to wall

4. Install crown molding

Interior partition

Movement takes place along this joint

edged, beveled, rounded, or molded at the bottom — against the wall, behind the crown molding. This additional trim improves the look of most simple moldings. When truss uplift is a concern, two pieces of flat trim will be needed. Start by installing the first piece of flat trim stock to the wall. Then, before installing the triangular nailing base, nail or screw the second piece of flat trim to the back of the triangular nailing base. Next, install the triangular nailing base by nailing it or screwing it to the ceiling.

Finally, the crown molding can be installed, using longer finish nails at the top of the crown and very short brads at the bottom. Since the crown molding is fastened to the ceiling and not the wall, some truss uplift movement can be accommodated between the two pieces of flat trim stock. All of the pieces except the triangular nailing base will need to be prefinished in order to allow movement to occur without revealing a line where the finish or paint stops. The crown molding will look best if the second piece of flat trim has a molded bottom — it will help disguise the unsecured sliding joint between the two pieces of flat stock.

— *Chuck Green*

Splices in Built-Up Beams

Q. Is it necessary to place splices in built-up lumber beams directly above the support posts?

A. The easy answer to this question is yes, but it's not entirely true. What is true is that locating all splices directly over support posts will keep you out of trouble. In reality, the most efficient location for splices is at points of inflection. The illustration below shows the expected deflection of a uniformly loaded beam without any splices spanning from wall to wall across a center post. Note how the beam sags near the centers of the spans, while the deflection curve turns upward over the post. The points where the curvature of the beam transitions from concave down over the post to concave up between the posts are the inflection points. At those points, stresses in the wood resulting from bending are lowest — in fact, they are zero.

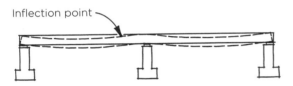

Inflection point

Bending stresses disappear at a beam's inflection points, making this a good place for splices in a built-up beam, as long as metal shear plates are used to handle shear stresses. But because it's difficult to figure out exactly where the inflection points are, it's always a safe bet to place splices directly over posts.

Unfortunately, shear stresses won't be zero at these points, so if you spliced all the members of a built-up lumber beam at inflection points, you would still need some type of steel or wood shear plates nailed or bolted across the splice to transfer the loads from one section of beam to the next. That's a trick that's common in commercial steel construction, but impractical for wood framing. A second problem is that wood beams aren't flexible enough to show the shape of the curvature and reveal the inflection points; their locations must be calculated. Since the location of each inflection point depends on the relative length of adjacent spans, the number of spans, and the variations in load along the beam, there is no easy rule of thumb for locating the inflection points and hence the best location for splices. So my suggestion is to take the safe route and set all your splices in multiple span built-up beams directly over the posts.

— *Christopher DeBlois*

Plywood in Built-Up Headers

Q. Do the layers of plywood in typical built-up headers add significant strength to the header?

A. The most important thing the plywood adds is thickness. Of course, the plywood does add some strength, but for several reasons engineers almost never count on this strength in their designs.

Only the layers of plywood with the grain oriented horizontally (parallel with the direction of the header) are really adding any strength. A quick look at the thicknesses involved shows that the additional strength is small. If half the layers in 1/2-inch plywood are horizontal, that's 1/4 inch of extra material. Compared with 3 inches of 2x10, that's an increase of only 8%. What's more, you only get the full effect of this extra thickness if there are no splices in the plywood near the middle of the span, or better yet, no splices at all. For headers at openings wider than 8 feet, that's not often the case. But it's these longer headers that will most likely need some extra strength.

Combine these drawbacks with size limitations and the plywood almost never makes a critical difference in safety. What I mean by size limitations is that when I design a header, the numbers may tell me I need two 2x9s. Since two 2x9s are about 30% stronger than two 2x8s, the 2x8s plus 8% from 1/2 inch of plywood wouldn't be strong enough. And I wouldn't ask the framer to rip some 2x9s, I'd simply call for 2x10s. What's more, he'll probably use double 2x10s for all his headers, big and small. Because headers only come in certain depths, there's usually extra strength in the 2x10s to begin with. And that extra strength in the 2x10s means that the small extra strength from the plywood is rarely important. But the thickness is helpful. — *Christopher DeBlois*

Nailing Patterns for Built-Up Beams

Q. What's the best nailing pattern for built-up beams?

A. The critical issue with built-up beams is that all the layers must deflect together and by the same distance in order to be properly sharing the load. For beams where the load comes down evenly on top of the beam, such as drop beams or beams directly under bearing walls, the nailing pattern is not all that critical. All you need are enough nails to hold the layers together and keep them from twisting. For beams loaded from the side, however, and especially for beams loaded from one side only, the nailing pattern is critical.

When beams are loaded from the side, such as a flush beam with joist hangers attached, there must be enough nails to transfer the load

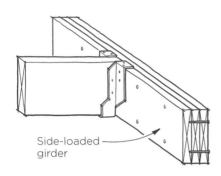

Side-loaded girder

Side-loaded beams must be carefully nailed together, to ensure that all the beam members share the load.

through the loaded member and into the attached members. For example, if a beam consists of three 2x10s loaded from one side only, the loaded member should carry only one-third of the weight. To transfer the rest of the load into the attached members, there must be enough nails from the loaded 2x10 into the center 2x10 to transfer two-thirds of the load, and enough nails from the far side 2x10 into the center 2x10 to transfer the final one-third of the load into that outer member. These numbers assume that all three 2x10s rest fully on the supports; the situation gets more complicated when the members are not all the same size or material. The bottom line, though, is that if all the pieces deflect together and equally, the beam should perform as designed.

At a minimum, for side-loaded beams, I recommend pairs of 16d nails every 12 inches along the beam, with the top row of nails $1^1/2$ inches or so from the top of the beam, and the bottom row $1^1/2$ inches or so up from the bottom. Use the same nailing pattern on both sides for triple beams, and check with an engineer whenever you think the loads involved might be unusually heavy.   — *Christopher DeBlois*

Using LVL to Reinforce Existing Beams

Q. Can laminated veneer lumber (LVL) be used to strengthen existing wood beams? I'll soon be working on a remodeling project where the existing floor system is supported by an undersized built-up wood beam, and I would like to stiffen the existing beam by bolting LVL material to it.

A. LVL can be used to strengthen or stiffen existing beams, but the manufacturer's load tables will generally not apply, and engineering analysis will likely be required to assess the member and connection capacities. The final design will depend on the geometry of the structural assembly, the design properties for the LVL, and the design properties of the existing framing.

In the case of a typical "top-loaded, simple-span" beam (Illustration A), the manufacturer's tables and installation guidelines may be used to size the LVL if the following conditions are met:

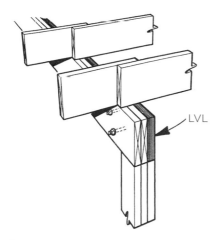
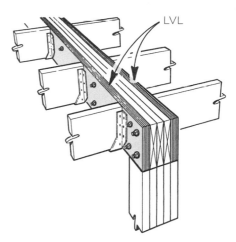

Illustration A. *In some instances, the manufacturer's span tables and installation guidelines can be used to size a retrofit LVL installation. The beam must be top-loaded, the LVL must be sized to carry the total load, and adequate bearing supports must be provided.*

Illustration B. *When supported members are framed into the side of the beam and/or the support reactions are provided by bolts rather than bearing, an engineer should size the LVL members and the bolt connections.*

- the supported members will bear on the top of the LVL;
- positive and adequate bearing supports are provided to support the beam-end reactions; and
- the LVL is sized to carry the total load on the assembly

In these situations, the bolts used to fasten the LVL to the existing beam generally have little structural demand and are only required to provide lateral support to the LVL members.

If any of the assumptions mentioned above are not true, as in side-loaded beams (Illustration B), then the bolted connections must be engineered based on the structural geometry and the design properties of the LVL and the existing framing. The bolts must be sized and spaced to direct the loads into the LVL along its length, ultimately reaching the end supports. Bolted connections (and the new and existing wood members) must be engineered according to the *National Design Specification for Wood Construction* (available from the American Wood Council, 800/890-7732; www.awc.org). When the existing framing shares in the load-carrying capacity, then the relative stiffness of the existing framing and the new LVL members must be evaluated to assess the proportion of the loads carried by each member. The design of the members and bolted connections for this situation can become very complicated and should only be done by an engineer.

— *Phil Westover*

Can You Rip LVL?

Q. The building codes don't allow you to rip graded lumber to a narrower width (to make a 2x6 and a 2x4 out of a 2x10, for example). But what about ripping LVL?

A. You're right about ripping graded lumber. Doing so "relocates" strength-reducing characteristics that affect performance, and effectively voids the grade designation (see illustration). It's different with LVL, which is a more consistent material. Manufacturers rip it as standard practice when producing beams. LVL is made in billets up to 48 inches wide that are ripped to commercially available standard depths. It's okay to rip LVL beams to a smaller depth, too, though you'll have to recalculate the beam capacity. Sizing software supplied by manufacturers such as TJ Beam (www.trusjoist.com) provides a sizing override where you can enter a non-standard depth to determine the load-carrying capacity of a ripped LVL.

— *Paul Fisette*

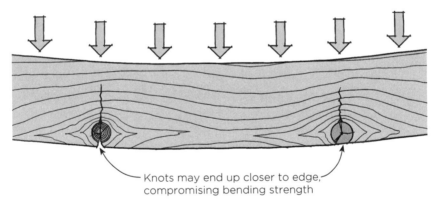

Knots may end up closer to edge, compromising bending strength

Ripping a piece of graded lumber to make a narrower beam is disallowed by code because it may place strength-reducing characteristics like knots close to the edge of the board, compromising bending strength.

Weight of Steel vs. Lumber Beams

Q. For the same loads, which is heavier, structural steel beams or lumber beams?

A. When depth is not a restriction, it is almost always possible to design a steel I-beam that is lighter than the lightest structurally acceptable wood beam design, including glulams, LVL, and Parallam beams. And no matter how hard you try, solid timbers, built-up 2-by beams, and flitch beams are almost always heavier than the lightest steel I-beam option — usually a lot heavier. Yes, it's true that steel as a material is heavier than wood given two chunks of the same size. That's because the density of steel is 12 times or so higher than the density of Southern Yellow Pine, for example. One cubic foot of steel weighs about 490 pounds, while the same size chunk of kiln-dried SYP wouldn't top 40 pounds. But because the steel can be formed into very efficient shapes, like I-beams, the overall weight of a steel beam is often lower than the lightest wood option.

In some cases, steel may be the only type of beam that will solve a problem. A good example is that common remodeling problem of removing a load-bearing wall without having the new support beam project below the ceiling. For long spans in a 2x10 floor, you can't get enough stiffness from 9 1/4-inch LVLs or 9-inch glulams, but 8-inch steel I-beams come in a variety of widths and weights to handle almost any situation like this. In such a case, the framer may complain that the steel beam is very heavy, but it's not heavier than the alternatives when there are none. There are also times when steel is ideal not because it can hold up a lot of weight, but because it can be welded into rigid frames. The modern two-story window wall leaves little room for plywood shear panels, but in high-wind and seismic areas you can't ignore the potential for racking that accompanies these lateral loads. A stiff moment frame of steel tubes or I-beams can often solve this problem when wood just won't do the job.   — *Christopher DeBlois*

Performance of Wood vs. Steel Beams in a Fire

Q. Is it true that a wood beam is safer than a steel beam in a fire? I've heard that metal twists and deflects in the presence of heat, while a wood beam will withstand the heat and a great deal of fire before burning through enough to collapse.

A. Large solid-sawn and glulam timbers provide a substantial degree of fire endurance. The superior fire performance of large timbers can be attributed to the charring effect of wood. As wood members are exposed to fire, an insulating char layer is formed that protects the core. Thus, beams and columns can be designed so that a sufficient cross-section of wood remains to sustain the design loads for the required duration of fire exposure.

A fire test conducted in 1961 at the Southwest Research Institute compared the fire endurance of a 7x21-inch glulam timber with a W16x40 steel beam. Both beams spanned approximately 43.5 feet and were loaded to full design load (approximately 12,450 lb.). After about 30 minutes, the steel beam deflected more than 35 inches and collapsed into the test furnace, ending the test. The wood beam deflected 2 1/4 inches with more than 75% of the original wood section undamaged. Calculation procedures provided in a publication from the American Wood Council, entitled *Technical Report 10: Calculating the Fire Resistance of Exposed Wood Members*, estimates that the failure time of the 7x21-inch wood beam would have exceeded 65 minutes if the test had not ended at 30 minutes. For additional information on the fire performance of wood, contact the American Wood Council at 202/463-4713 or www.awc.org. — *Brad Douglas*

Strength of Built-Up vs. Solid Lumber Columns

Q. When I need a 6x6 deck post, I usually assemble a built-up column from three 2x6s. Is this kind of post equal in strength to a 6x6?

A. The strength of the built-up post you describe depends on the grade and species of the lumber, and how the 2x6s are nailed together. Assuming the post and 2x6s are No. 2 Southern Pine, the nailing used to laminate the three 2x6s is the main issue in determining if three pieces of 2x6 will replace the 6x6 post.

When the deck column is loaded, the nails or screws slip or give slightly, allowing the column to start bowing. If the nailing is inadequate, the bowing will continue and the capacity of the deck column is reduced. Even when such a column is assembled in accordance with rigorous National Design Specification for Wood Construction (NDS) provisions (using 30d common nails 8 inches on-center and staggered $2^{1}/2$ inches), its bearing capacity is reduced 40% from a solid-sawn column of the same size ($1^{1}/2$ x $5^{1}/2$ inches) having identical lumber properties.

In summary, substituting a post made of laminated 2-by material for a solid-sawn post in a freestanding application (such as a deck post) is not recommended without an in-depth analysis based on the NDS column provisions. — *Frank Woeste*

Exterior Stair Riser Heights

Q. I'm building an exterior stairway from a deck to grade. Is there a building code requirement that the risers of an exterior stairway all be the same height?

A. Most building codes, including the International Residential Code (R311.5.2), require any flight of stairs to have no more than $3/8$-inch difference between the largest and smallest riser, and the same rule holds for treads. There are a couple of exceptions: 1) where the bottom riser adjoins a sloping public way, walk, or driveway that has an established grade and serves as a landing; and 2) on stairways serving as aisles in assembly seating. In those cases, the bottom riser may be wedge-shaped if necessary.

As far as I know, there is no reason to think that an inspector would treat exterior stairs differently from interior stairs with regard to dimensional uniformity. — *Paul Fisette*

Durability of Cedars

Q. I'm pricing cedar for a replacement deck that was formerly built with treated yellow pine. A local supplier is telling me his "northern white" cedar will last far longer than "Alaskan" cedar, which I can purchase for about half the price. Is there really such a dramatic difference?

A. According to all my reference materials and based on my own experience, Alaskan, northern, Atlantic, and Port-Orford cedar are similar in durability. Keep in mind that only the heartwood is resistant to decay. Sapwood is not at all reliably durable. One difference with Alaskan yellow is that the color of the heartwood is yellow, making it easier to distinguish the heartwood from the sapwood (which is pale yellow). With the other cedars, there is not a great difference between sapwood and heartwood colors, so it can be more difficult to tell if you have in fact purchased durable heartwood or nondurable sapwood. — *Paul Fisette*

Durable Exterior Ceilings

Q. Patio covers are standard features of nearly every home built in the Phoenix area. Standard construction is wood frame with a drywall ceiling, taped, textured, and painted. Within 10 years, the taping and texturing start to separate from the drywall. Retaping and texturing might make it last a few more years. Is there an economical alternative to drywall that would provide a more permanent and relatively maintenance-free ceiling?

A. Drywall is not designed for exterior applications, even in low-moisture desert climates. Although it will cost more, the best solution for a patio cover such as you describe is to use a water-resistant cement board (like USG's Durock) or gypsum fiber panel (like USG's Fiberock) instead of the drywall. After attaching the cement board or gypsum fiber panels to the framing, cover the joints and the panel surface with a latex-modified thinset base coat. You can then finish the surface with an aggregated acrylic textured finish to provide a stucco look. These base coat and finish materials are available from manufacturers of exterior insulation and finish systems (EIFS), such as Dryvit, Senergy, Parex, and Sto. — *Jim Reicherts*

Porch Floor Board Orientation

Q. I'm building a roofed porch with tongue-and-groove fir decking in New England. The boards will be primed and painted on all sides before installation. The joists run perpendicular to the building, and I'd like to install the flooring parallel to the long dimension of the porch. Does the orientation of the porch boards matter?

A. The drainage off the deck will definitely be better if the boards run from the house down the slope toward the edge of the porch. I usually slope my porch decks about 1/4 or 3/8 inch per foot. I've found that 1/8 inch per foot is not enough to get the water moving off the porch.

If you are concerned about end checks developing in the exposed end grain of the deck boards, I suggest treating the end grain of the boards with penetrating epoxy consolidant before any other finish is applied. This seals the end grain, limiting the movement of water in and out and minimizing checks. — *John Leeke*

Oozing Knotholes in PT Deck

Q. Is there anything that can be done, short of tear-out and replacement, to prevent the knotholes in pressure-treated pine decking from oozing on hot days?

A. The sap in the knotholes softens and begins to run as it is heated by the sun. Sap is composed of liquid and solid materials. If it is heated to 160°F while it is being kiln-dried, the volatile liquid substances will flash off, leaving the solids behind in a hardened state. Once the sap has set, it will never run again. This doesn't help you now, however, since the wood in your deck was obviously not kiln-dried at a high enough temperature during production.

There are a couple of options. You can wait it out; sooner or later the sap will stop running out of the wood. In the meantime, you can clean it up with turpentine. Or you can try to set the sap in the knots using a heat gun. Just be careful not to start a fire or singe the wood. — *Paul Fisette*

Bonding Treated Wood

Q. What kind of adhesive should I use to bond CCA-treated decking to treated joists?

A. Preferably none. The U.S. Forest Products Laboratory does not recommend any adhesive between the decking and the joists. The swelling and shrinking of the two members at right angles to each other will break the bond of any rigid adhesive. Many of the more flexible mastic-type products on the market will stick to treated wood, but the differential movement will cause failure in a few years. Neither type of adhesive is strong enough to prevent deck board cupping. — *Henry Spies*

Should Deck Posts Go in Ground?

Q. Can 6x6 pressure-treated deck posts be sunk into the ground or should they be supported on concrete piers?

A. Theoretically, posts that have been pressure-treated for ground contact can be sunk into the ground. However, it is unlikely that the treatment has penetrated fully through a 6x6. So I would opt for the concrete piers with the posts supported on metal post anchors.

If the posts are sunk directly into the ground, the bottom of the post must not be cut. I have seen a number of posts with the center rotted out because the treatment chemicals did not penetrate completely and water was wicked up through the untreated center. Post tops should be beveled to shed water or have a metal post cap added. Any cut ends of pressure-treated material that are left exposed to the weather should be dipped in or painted with a copper napthenate solution. — *Henry Spies*

Penetration of PT Chemicals

Q. Our local lumberyard carries .40 CCA-pressure-treated wood. The label says the wood may be used for ground contact. The wood has needle marks from the treating process. When I cut the lumber, the treatment appears to be only about 1/4 inch deep, which doesn't seem adequate for sill plates in contact with concrete. I have also used .60 PT, and it shows green all the way through when cut. However, my supplier assures me that the .40 PT will not fail from either insects or moisture. Is he right?

A. How deeply the chemicals penetrate the wood during treatment is indeed important. Penetration levels vary widely. Heartwood is more difficult than sapwood to treat. The heartwood of Douglas fir and southern pine (both commonly used for pressure treatment) resists penetration and may allow only 1/4 inch of chemical penetration. So if the lumber you purchased was heartwood, it is not surprising that you would see shallow penetration. The good news is that the heartwood is typically more rot resistant than the sapwood. The heartwood of both Douglas fir and southern pine is considered moderately decay resistant, but virtually all treated southern pine is second growth, easy-to-treat sapwood.

The 0.40 designation of the wood you bought means that the amount of chemical retained by the wood after treatment (its retention level) is roughly equal to 0.40 pounds of chemical per cubic foot of wood. That is the correct amount for "ground contact." The "needle marks" that you see are a result of incising, a process in which lumber is passed through a series of rollers equipped with teeth that sink about 1/2 inch into the wood. The incisions expose the more absorbent end grain of the wood throughout its length, allowing better penetration and chemical retention. Typically, incising is used for more difficult-to-treat species like Douglas fir (which can have more heartwood) and not southern pine.

So is your wood okay? Probably. There's a good chance it's heartwood, and even the shallow penetration of heartwood afforded by incising has some value. Rot fungi usually begin to grow not in the middle of the wood but on the surface. Where there are cracks or checks in the wood, incising helps reduce the likelihood that fungi can get into those pathways and rot the wood from within. — *Paul Fisette*

Box, Sinker, and Cooler Nails

Q. I would like to know the definitions for a box nail, sinker nail, and cooler nail. How do they differ from common nails?

A. A box nail has a smaller gauge shank than a common nail. For example, while a 10d (3-inch-long) common nail has a 9-gauge (0.148-inch-diameter) shank, a 10d box nail has a $10^1/2$-gauge (0.125-inch-diameter) shank, which is thinner. Because a box nail is lighter than a common nail, it has less shear strength.

Cooler and sinker nails also have a smaller-gauge shank than a common nail but not as small as the shank of a box nail. 10d sinkers and 10d coolers are both a little shorter ($2^7/8$ inches), and have an 11-gauge (0.120-inch-diameter) shank. Sinker and cooler nails are usually resin-coated; the resin acts as a lubricant as the nail is driven and as an adhesive when the nail cools after being driven.

Cooler nails, like common and box nails, have a flat head. Sinker nails, on the other hand, have a bugle-shaped head to promote countersinking.

All of these terms — common, box, sinker, and cooler — refer to loose nails, not collated nails. Collated nails are categorized according to head type (full round head or clipped head), length (in inches), and shank diameter (in inches). For example, a 16d common nail (loose) is equivalent to a full-round-head, $3^1/2$-inch-long, 0.162-inch-diameter collated nail. — *Scott Smith*

Clipped-Head vs. Full-Head Nails

Q. What are the pros and cons of clipped-head versus full-head nails? Are there any framing nailers that drive both kinds?

A. Clipped-head nails are collated at a steeper angle than round-head nails. The steeper collation angle provides two main advantages: there are more nails in a stick, and the working clearance (the angle between the gun and the work) is greatly improved.

The model building codes do not require the use of round-head nails. The codes specify nails by length and by shank diameter, not by the type of nail head. Lateral load capacity depends on a nail shaft's shear strength, and withdrawal resistance depends on shank type. An increase in withdrawal resistance can be achieved by using a screw-shank or ring-shank nail.

The industry reference for evaluating pneumatic nails is the National Evaluation Report (NER-272), which is posted at the Senco Website (www.senco.com/pdf/facts/ner272.pdf). This report lists the nailing schedules for pneumatic fasteners for all of the model building codes (BOCA, CABO, SBCCI, and ICBO).

A panel's shear strength depends more on the depth to which fasteners are driven than on the shape of the nail head. Nail heads should be set flush with the surface of the sheathing, not countersunk. When fasteners are driven through the outer ply of plywood or OSB, shear strength decreases significantly. But if the nails are properly driven, there is no difference in performance between the two types of nail heads, because the pull-through values of both nails exceed the performance requirements for the assembly. Of course, regardless of the type used, proper spacing of the nails is essential.

Although there is no evidence that round-head nails provide greater shear strength or withdrawal resistance than clipped-head nails, some building inspectors ignore the facts and require round-head nails on exterior sheathing. These inspectors reason that the larger head area of round-head nails reduces the likelihood of overdrive. Currently, all manufacturers make separate nailers for the two types of nail heads, and the nails are not interchangeable. Nails with a third type of head, the offset round head, are available from Paslode, Senco, and Hitachi. These nails can be used in clipped-head guns.   — *Eric Borden*

Purpose of T&G Plywood

Q. Does tongue-and-groove plywood add extra strength or stiffness to a floor system, or does it just help prevent floor squeaks?

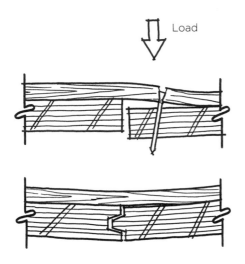

Unsupported joints in square-edged plywood floor sheathing cause squeaks when someone steps on them. T&G plywood sheathing prevents squeaks and makes the floor feel stiffer, though it doesn't actually increase the design strength of the floor system.

A. The tongue-and-groove joint doesn't add strength, but it does help to distribute loads to adjacent panels, improving the perceived stiffness of the floor. T&G plywood was developed as a labor-saving alternative to installing solid wood blocking at unsupported panel edges. Without the tongue and groove, a load on one panel edge causes that panel to deflect relative to the adjacent sheet. A wood floor that spans across the joint would experience a wedging action, causing a floor squeak. Tongue-and-groove plywood is actually more effective than solid blocking at preventing squeaks, because over time the blocking will shrink, leaving unsupported edges. — *Scott McVicker*

Gaps Between Sheathing Panels

Q. Most sheathing manufacturers recommend that panels be spaced at the edges and ends. Since panels measure a full 4x8 feet, the only way I've been able to provide this spacing (and maintain standard joist and stud layout) is by trimming the ends and edges of the sheathing panels. Why don't manufacturers size the panels 47⅞ by 95⅞ inches?

A. The APA, as well as most panel manufacturers, recommends $1/8$-inch spacing at the edge and end of panels. APA manufacturing standards are designed to accommodate this spacing by permitting a full $1/8$-inch plus-or-minus tolerance on the length and width of sheathing panels. Many manufacturers do, in fact, cut their panels a little short and narrow. While the amount of cutback varies among manufacturers, it's typically $1/16$ inch. Why not a full $1/8$ inch? The $1/16$ inch may be a compromise to avert rejected panels that are too short or out of tolerance.   — Ed Keith

Swelling of OSB Roof Panels

Q. We build timber-frame homes and wrap them with insulated panels made with skins of OSB on the outside and rigid foam insulation on the inside. We've been pleased with the results except for one thing: the swelling of the OSB at the edges. We've seen no damage to the roof structures, but the ridging is unsightly when the roofs are covered with asphalt shingles. So far, we haven't been able to find a solution, other than to make sure that the roofer covers the roof immediately with felt and uses a good-quality shingle. Have you seen any instance where this swelling has contributed to roof failure?

A. The problem, caused by moisture adsorption, is fairly common with stressed-skin panels installed on roofs. The panel edges take on moisture more quickly than the rest of the panel and swell as a result — one of the shortcomings of OSB. What can happen with roof panels is that warm, moist air from inside the house leaks through the seams. As the indoor air cools, the moisture condenses and wets the panel edges. The moisture may also collect on the underside of the impermeable asphalt roof felt and shingles, compounding the problem. The solution is to meticulously seal the seams of adjoining panels with a product like spray urethane. Inject foam-in-place urethane into the seams before assembly or drill holes and fill the seams after the panels have been installed. Be careful about using spray-in-place urethane to seal panels in cold weather. The necessary heat of reaction is wicked away by the cold, which can inter-fere with the curing process and leave an unprotected seam. Check with the spray-foam manufacturer for recommendations on cold-weather installations.   — Paul Fisette

Deflection of Plywood vs. OSB

Q. Does OSB sag more than plywood when installed horizontally over 24-inch-center rafters?

A. It depends on the materials used to make the OSB, which can be manufactured from a variety of species. These include aspen, southern pine, sweet-gum, yellow poplar, and birch. The Modulus of Elasticity (MOE) of the wood used will determine the relative flexibility of an OSB panel. The list below shows the MOE values for some of the woods used to make OSB and plywood (from NDS Supplement). On the same roof with rafters spaced at 24-inch centers, a plywood panel made from high-grade Douglas Fir-Larch veneers is going to deflect less than an OSB panel made from Aspen. Run the same test using an OSB panel made from similar materials and you will

Wood Species	MOE
Aspen	800,000 to 1,100,000
Yellow Poplar	1,100,000 to 1,500,000
Beech-Birch-Hickory	1,200,000 to 1,700,000
Douglas Fir-Larch	1,300,000 to 1,900,000
Southern Pine	1,200,000 to 1,900,000

most likely find no difference in their deflections. Probably the reason that OSB has the reputation for flexing more than plywood is that much of the OSB sold is manufactured from lower-grade fibers. This is why OSB typically costs a lot less than good plywood. — *Scott McVicker*

Spacing of T&G Plywood

Q. How should tongue-and-groove plywood be laid? Do you have to account for expansion between sheets?

A. The American Plywood Association recommends that T&G plywood sheets be spaced 1/16 inch apart at both sides and ends. The thickness of a nickel is about right. A little more space is acceptable; less is not. This applies whether or not a construction adhesive is applied to the joint. Many panels these days have a longer tongue than a groove, so the spacing is automatic. — *Henry Spies*

Fire Ratings for Engineered Lumber

Q. Do laminated veneer lumber (LVL) and wood I-beams have a lower fire rating than conventional wood framing members? If so, have the model codes addressed this?

A. Individual LVL members or wood I-beams won't resist fire as well as the solid wood members they replace. However, fire ratings are based on assemblies, not individual members. Several manufacturers of LVL and wood I-beams have obtained fire ratings on whole floor and ceiling assemblies by using their products with a Type X drywall facing the interior. For example, a one-hour fire-resistant ceiling assembly using TJI joists requires two layers of 1/2-inch Type X drywall applied to the bottom joist flange. A comparable one-hour assembly using solid-sawn joists requires one layer of 5/8-inch Type X drywall. Since there are no "generic" wood I-beams (each is engineered to perform in specific ways), there is no generic assembly that achieves a certain fire rating. Each manufactured product and assembly must be approved after testing by a certified laboratory. Check with the manufacturer for each product you use. There are several small manufacturers whose volume does not justify the expense of obtaining a fire rating, and their products cannot be used where a fire rating is required. *— Henry Spies*

Engineered Lumber Stringers

Q. Can I use laminated veneer lumber (LVL) for stair carriage material? Since it doesn't shrink, it seems it would solve many of the problems associated with the shrinkage of sawn lumber stringers.

A. LVL should work fine for stair carriages in most cases, but you should check with the manufacturer before cutting. Because LVL is intended primarily for use as a beam, manufacturers often warn against notching. Otherwise, it has not only the advantage you mention, but also superior strength. And because LVL is available in wider sizes than dimensional lumber, it may be useful, if sized correctly, for longer stair spans than sawn stringers can safely handle. The engineered lumber manufacturer can help you select the best product for a stair application. For example, in addition to Microllam LVL, Trus Joist also makes Timberstrand LSL (laminated strand lumber). LSL is less expensive than LVL, is easy to cut, and has good resistance to nail splitting. To assist builders, Trus Joist offers a stringer sizing chart that specifies minimum throat depths and allowable stair spans for Timberstrand stringers. For a copy, call 800/628-3997 or go to www.trusjoist.com.   — *Curtis Eck*

LVL Dimensions

Q. Why are LVLs sized differently than framing lumber? I can understand the $1^3/4$-inch-thickness dimension, because two laminations make up a matching header for a 2x4 wall, but why are they $9^1/2$ inches deep instead of $9^1/4$?

A. First of all, LVL is also sold in depths of $7^1/4$, $9^1/4$, and $11^1/4$ inches to match standard stick-framing sizes. Take a look at the LP Gang-Lam site (www.louisianapacific.com) or the Trus Joist Microllam site (www.trusjoist.com), for example. However, LVL is also available in non-standard sizes like $9^1/2$ and $11^7/8$ inches all the way up to 24 inches deep. You will find that the deeper $9^1/2$-inch and $11^7/8$-inch versions carry a little more load than the $9^1/4$-inch and $11^1/4$-inch stock, but that's not why distributors push nonstandard depths. It's actually because the manufacturers don't want you to mix engineered wood with sawn lumber.

Look at the product literature for I-joists, which are made by the same companies that sell LVL. You will see that they sell $9^1/2$-inch and $11^7/8$-inch I-joists as well, not $9^1/4$- or $11^1/4$-inch. Swelling, shrinkage, and stability characteristics of engineered lumber are much different than those of sawn lumber, so mixing is discouraged. That's why there's been a trend toward engineered-wood floor systems, which use a combination of LVLs and I-joists but no dimension lumber. Nevertheless, manufacturers realize that builders will have projects where matching sawn lumber is required, especially in remodeling, so they also offer "standard" sizes.   — *Paul Fisette*

Roofing

3

Wind pressure overcomes buoyancy of warm air

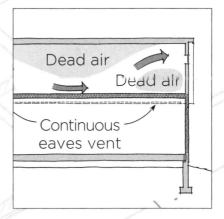

Dead air

Dead air

Continuous eaves vent

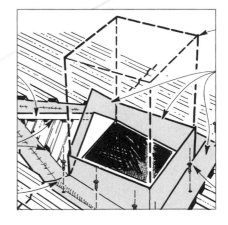

Will a Steep Roof Last Longer?

Q. Is it true that the steeper a roof is, the longer asphalt shingles will last?

A. Yes, the steeper the roof, the longer any roofing material will last. Because water runs off a steeper roof more easily, the roof covering will dry faster. And during the hottest part of the year when the sun is higher, a steeper roof will absorb less solar radiation than a flatter roof. There are limits, however, since some materials will not be heavy enough nor the adhesive strong enough to hold the roofing material onto very steep roofs, such as those greater than 20:12. Also, at very steep slopes, granules may wear out faster in the cutouts of traditional three-tab shingles. Fortunately, not a lot of designs call for such slopes. — *Henry Spies*

Why Paper a Roof?

Q. What is the purpose of tar-papering a roof, other than to keep water out until the shingles are laid down? It seems that if water does get past the shingles, then the tar paper, with its many staple and nail holes, would do little to keep the water out.

A. Tar paper, or roof felt, is required by most building codes. Roof shingle manufacturers require it to maintain the warranty on the shingles. And, as you point out, roof felt provides temporary protection from the weather while you close in the house. Also, the fire rating of a roof covering is assigned to an entire roof assembly, including roof felt. If you don't install the felt, you don't earn the fire classification that may be required by the local code.

In addition, applying roof felt can reduce telegraphing of sheathing seams through thin asphalt roof shingles. However, if you install the felt carelessly, it can create a lumpy look.

Roof felt does provide a long-term benefit — a second line of defense if water gets past the roof covering. Strong wind can blow rainwater in along the edges of the roof or even lift off shingles. True, the roof felt does have nail penetrations and water can leak through these nail holes. But the holes are small and fasteners do not penetrate most of the roof surface, so the membrane can keep a lot of damaging water out of the roof frame. If there's no roof felt and you develop a leak, the water will run down under the shingles and pour through the seams in the roof sheathing.

Overall, water intrusion is likely to be much less with felt than without. It may buy you time to repair the roof before any major damage is done to the structure. — *Paul Fisette*

Asphalt Shingles & High Wind

Q. Is the self-adhesive strip on standard asphalt shingles adequate for occasional 100-mph winds, or should extra roofing cement be applied under each shingle?

A. Wind-resistant shingles are recommended in areas subject to hurricane-force winds, which are defined as winds in excess of 75 mph. Typical asphalt roof shingles have a UL 997 rating and come with a "wind resistant" label right on the bundles. However, this UL 997 listing only tells you that the wind testing was conducted in a certain way; it doesn't refer to a specific wind-speed resistance. In fact, most shingle manufacturers limit their warranties to wind speeds between 60 and 80 mph.

Because you are in an area subject to even stronger winds, you may want to take extra measures. Shingle manufacturers recommend that in windy locations you place a dab of asphalt cement the size of a quarter at the bottom corner of each tab of a standard three-tab shingle (a total of six dabs). Double-nailing (for a total of six nails per shingle) is also required by code in high-wind zones and specified to uphold the warranty on wind-resistant shingles. And it's best, if possible, to install the shingles in warm weather so the self-sealing mechanism works properly.
— *Paul Fisette*

Shingle Overhang

Q. How far beyond the drip-edge should an asphalt shingle extend? Is the amount of overhang the same at the rake as at the eaves?

A. The Asphalt Roofing Manufacturers Association's *Residential Asphalt Roofing Manual* calls for a 1/4-inch to 3/4-inch overhang at both rakes and eaves, and permits flush-cutting with drip-edge. However, most manufacturers recommend a 1/4- to 1/2-inch overhang at the eaves and rake with a drip-edge. In my opinion, the felt underlayment should also extend beyond the drip-edge (by $1/4$ to $1/2$ inch). Don't overdo the overhang: shingles that extend more than $3/4$ inch can eventually bend and fracture, leaving the roof sheathing and fascia vulnerable to water damage.

— *Harrison McCampbell*

Asphalt Shingle Nailing

Q. Can asphalt shingles be nailed *above* the adhesive strip? I have seen leaking roofs with relatively new asphalt shingles where the leaks seemed to follow the nailing pattern. In the attic, there were rows of drip holes in the cellulose insulation on the attic floor. In this case, the shingles were nailed below the line of the self-seal

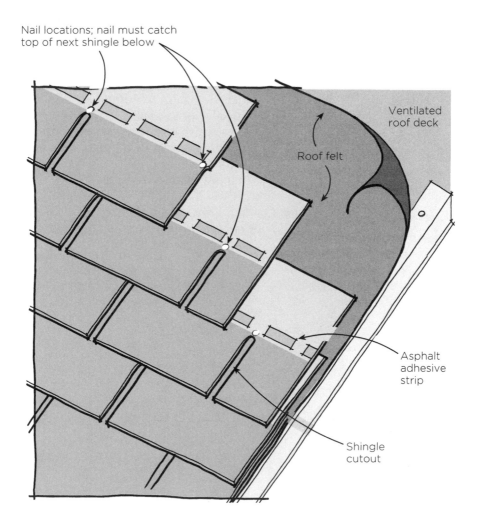

Nail locations; nail must catch top of next shingle below

Ventilated roof deck

Roof felt

Asphalt adhesive strip

Shingle cutout

adhesive, according to the manufacturer's instructions. (We use fiberglass shingles with an adhesive strip.) Wouldn't it be better to nail above the adhesive, where no rain can blow up to get at the nail heads?

A. Asphalt shingles should always be installed according to the manufacturer's instructions, which are usually printed on the shingle wrapper. Because the thickness, weight, and shape of shingles vary, manufacturer's instructions vary slightly. But virtually all manufacturers agree that nails should be installed below the adhesive strip. To keep a warranty intact, follow the manufacturer's nailing and installation instructions closely. If you deviate, whether you think it is a better installation or not, be prepared to roll with the directive from the manufacturer if there is a leak or performance question.

If you nail too low, nail heads can be exposed, with the potential for water entry. If you nail within the adhesive strip, the nail head cuts down on the amount of asphalt surface available to hold the shingle down. If you nail too high (above the adhesive strip), the nails may miss the top of the shingle underneath, failing to support the underlying shingle and possibly causing a tear or hump in the upper shingle. Most people don't realize that each nail holds down two shingles at once. If the nail is not at least 1/4 to 1/2 inch below the top of the shingle below, it may tear through. Proper nailing is especially important with the new "dimensional" (thicker and heavier) shingles, which, if not nailed properly, have a tendency to slide.

In the case of the dripping nails, it is more likely that the drips were caused by condensation, not by roof leaks. In cold climates, if the roof surface is very cold and the air in the attic is relatively warm and humid, condensation can form on the underside of the roof sheathing and drip from projecting nail points. If the low-slope roof is actually leaking, the installation may have been incorrect. All shingle roofs should include a layer of unperforated organic (not fiberglass-based) felt underlayment. Shingles should not be installed on slopes of less than 4:12, unless specific issues have been addressed (see "Installing Asphalt Shingles on a Low-Slope Roof," below). This is both a code requirement and a requirement of most shingle manufacturers. — *Harrison McCampbell*

Installing Asphalt Shingles on a Low-Slope Roof

Q. What is the minimum slope allowed for asphalt shingles? I have heard that asphalt shingles can be applied on a roof with a pitch as low as 2:12, as long as a rubberized asphalt eaves membrane is installed under the shingles.

A. The International Residential Code permits the installation of asphalt shingles on roofs with slopes as low as 2:12 to 4:12 if the shingles are double-coverage, self-sealing versions. Shingles that are not self-sealing must be hand-sealed with asphalt roofing cement. Low-slope shingle application requires that the No. 15 felt underlayment be installed with 19-inch top laps and 12-inch side laps to provide double coverage. In areas where the average January temperature is 25°F or less, low-slope applications require an eaves flashing consisting of two layers of No. 15 felt underlayment, cemented together, for a distance extending from the eaves up to a point 24 inches inside the interior wall line of the building.

Asphalt shingle manufacturers will usually warrant shingles installed on roof slopes as low as 2:12 when these special installation procedures are followed. Self-sticking bituminous membranes like Grace Ice & Water Shield are acceptable for both the underlayment and eaves flashing.

Having said this, I would never install shingles on a 2:12 roof, especially in an area where snow falls. In fact, I have had numerous problems in northern climates with asphalt shingles on 3:12 roofs. As a result of this experience, I would limit the application of asphalt roof shingles to roofs that are 4:12 and steeper. The National Roofing Contractors Association recommends that asphalt shingles be installed only on roofs with slopes that are 3:12 and greater.

I believe that if you have to depend on felt paper, asphalt cement, or bituminous membranes to block water after it gets past the primary roof covering, you have the wrong roof system. Shingles get brittle with time, and at the reduced slope of 2:12, a slight curl at the end of the shingle line shunts water backward under the leading edge of the overlapping shingle. When you have snow or ice sitting on the roof and melting, you don't have a prayer at keeping the water out. Felt paper and bituminous membranes are emergency backups, not a plan for everyday protection.

— Paul Fisette

Asphalt Shingles on Unvented Roofs

Q. Our insulation subcontractor advises us to install dense-pack cellulose in the rafter bays over a cathedral ceiling, without ventilation baffles, and the local building inspector approves. If we don't ventilate under the roof sheathing, what happens to our asphalt shingle warranty?

A. With few exceptions, installing shingles on an unventilated roof deck will void the warranty. The fact that your local building department okayed the installation probably won't help: As the warranty from the Canadian firm EMCO, which makes Esgard shingles, states, "Where local building codes have specific [ventilation] requirements that differ from the National Building Codes, the more stringent requirement must be followed." (EMCO: 800/567-2726, www.emcobp.com)

In fact, as you might guess from that quote, just having the roof ventilated may not be good enough. The shingle maker has to agree that the ventilation is up to its standards. If the warranty excludes installations with "improper ventilation," that language may be enough to deny a claim.

Unfortunately, shingle warranties provide little protection, whether or not your roof deck is ventilated. Almost no roof is put on perfectly, and if you've deviated from the manufacturer's instructions in any way, your claim can be denied. In the end, it comes down to trust: If manufacturers want to back the product up, they will, and if they don't, they won't. You've got to decide if you want to trust them. And even if a warranty is honored, the money you'll get will likely be prorated and will not cover costs like tear-off, disposal, or labor.

The real question is, how will the shingles hold up on an unventilated roof deck? They might give out a little sooner than shingles on a ventilated roof (maybe by a year or two), but probably not enough to notice. The most important factor in longevity isn't the level of ventilation; it's the quality of the shingle. The better brands of fiberglass shingles are the ones labeled as passing ASTM Standard D-3462. If you start with a good shingle, what you lose in shingle life (if anything) will probably be more than earned back in energy savings from the added insulation thickness. At least one manufacturer, CertainTeed, will honor its warranty on an unventilated roof deck, although for a reduced term of 10 years (prorated from year one, and with no wind-speed rating). CertainTeed has funded a lot of research on the causes of shingle failure, including some long-term studies at university sites in three different climates. In these studies, shingles were applied on ventilated and unventilated cathedral roofs side by side. The research indicates that high temperatures do cut the life span of shingles, but only marginally; and it shows that roof ventilation doesn't have much of an effect on shingle temperature anyway (shingle color and roof orientation are more important). However, ventilation affects not only shingle temperature but also the level of moisture in the roof assembly, as well as the melting and refreezing of snow on the roof. Because these factors can affect how a shingle ages, it may be reasonable for manufacturers to limit or exclude warranty coverage on unventilated roofs.

— *Ted Cushman*

Reroofing Over Asphalt Shingles

Q. Are there any drawbacks to reroofing over asphalt shingles? How many layers can accumulate before you should tear them off?

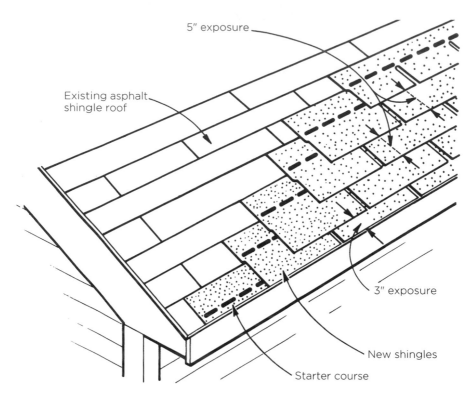

5" exposure

Existing asphalt shingle roof

3" exposure

New shingles

Starter course

When installing a second layer of asphalt shingles, cut the tabs off the starter course and butt each course against the end of the old tabs. This will keep the roof surface smooth.

A. There are no problems with adding a second layer of shingles over conventional asphalt shingles if they are properly installed with nails that fully penetrate the roof sheathing below. To keep the second layer flat and avoid a wavy pattern in the new roof, cut the tabs off the new shingles for the starter course, as shown in the illustration. Nails, not staples, should be used to fasten the new shingles.

I would not recommend more than two layers of shingles, although some codes permit three under some circumstances. There are three reasons for not going past two layers:

1) the combined weight of the layered shingles would be very heavy and might exceed the design load of the roof structure;
2) the deep layers would give inadequate lateral support for the nails and there would be a greater chance of a blowoff; and
3) the roof would retain more heat, which could reduce the life expectancy of the shingles.

I would also not recommend a second layer over textured (architectural-style) asphalt roof shingles, because of the added weight and the difficulty of producing an even surface. —*Henry Spies*

Removing and Preventing Asphalt Shingle Stains

Q. In this area of eastern Virginia, which is rainy and humid in the spring, hot and humid in the summer, and warm and humid in the early fall, black staining on asphalt shingles is common. This staining appears to be caused by a mildew or fungus.

I have made several observations about the problem in this area:

1) Affected roofs are of a light color (perhaps the problem just doesn't show on darker roofs);
2) Affected roofs are in areas with trees nearby, but not necessarily shaded;
3) It appears on older roofs, perhaps more than ten years old, but the roofs do not appear to be deteriorated;
4) Staining is more common on lower parts of roofs than upper parts;
5) Roof pitch does not seem to be a factor;
6) Not all roofs of the same age and color have the same problem.

I have asked several roofing contractors about cleaning off the stains, but none have tried. How can the stains be removed or prevented?

A. The discoloration is actually caused by a pigment produced by a genus of blue-green algae known as *Gloeocapsa*. It appears throughout the U.S. in warm, humid climates, but it thrives in the Gulf states, along the eastern seaboard, and in the Northwest. It exists on darker roofing, too, as you guessed it might, but the stain is not as visible as it is on white and pastel roofs. Despite its unsightly appearance, algae growth will not affect the durability of the shingles.

Algae discoloration is difficult to remove from roofing, but it may be lightened with a solution of one part chlorine bleach to three parts water, with some trisodium phosphate added. This treatment will lighten the staining, but it will not completely remove or prevent the algae from returning.

The solution should be applied with a soft brush to avoid dislodging the granules. During application, this solution will make the roof slippery and hazardous, so work from ladders or scaffolding. The roof should be rinsed with a gentle spray from a hose after 20 to 30 minutes. Be aware that the solution and rinse water will kill grass and landscaping plants, so the water from downspouts should be collected and the landscaping covered with plastic during the treatment. Rinse the gutters and downspouts, as well.

Woodstream Corporation (800/800-1819, www.saferbrand.com) makes a product called Safer Moss and Algae Killer, made for exterior use. I have had no experience with it, but it is advertised as being biodegradable and should make the issue of getting the solution on lawns and gardens less of a problem.

New "fungus-resistant" shingles are available in most parts of the country where this algae thrives. These shingles have zinc mixed into the ceramic-coated granules. The zinc coating slowly dissolves, and the metal prevents algae growth. Any of the heavy metals, such as lead and copper, will work, but zinc seems to be the most economical. The effect of the metal is dramatically visible on some roofs, where the areas below the lead, zinc, or copper flashings are unstained, while the rest of the roof is darkened by the algae. The scattering of gray granules in the shingles is not noticeable in most shingle colors and patterns. Of course, these shingles are more expensive because of the cost of the zinc granules.

To prevent staining on existing roofs, try slipping a narrow strip of zinc under the ridge row of shingles, so about 3/4 inch is exposed. A second strip can be installed about halfway down the roof. This will have essentially the same effect as the zinc granules. A similar installation of a strip of copper will control moss and mildew growth on wood shingles.   — *Henry Spies*

Metal Roofing

LIfe Expectancy of Metal Roofs

Q. We're working on a 120-year-old house that still has its original standing-seam metal roof. A roof painter told the owner that this roof has 50 years of useful life left. Can a metal roof really last 170 years?

A. If this roof is 120 years old, that would mean it was built around 1880. In those days, the options were copper, lead, tin-coated iron, and terne-coated steel. Tin-coated malleable iron was disappearing at the time. Copper and terne rolled roofs were very popular during that vintage — terne more so because it was less expensive. Terne is an alloy of lead and tin that provides excellent corrosion protection for steel. It was recently taken off the market due to the health concerns about lead, although no specific health threat was ever established. The replacement material, manufactured by Follansbee Steel, is a zinc-tin alloy called Terne II (800/624-6906, www.follansbeeroofing.com).

Regarding this particular roof, it would not surprise me to see a copper, lead, or lead-coated copper roof last 150 or even 170 years. There are many examples in Europe, and a few here in the States, that are even older. But since you mention a roof painter, it seems likely that this is a terne roof, not copper or lead. Lead and copper roofs are rarely painted; terne steel roofs must be cleaned and painted periodically. If they are kept up that way, however, terne roofs also can last a very long time. A lot of terne roofs are a good 100 years old. To say a terne roof would last 170 years might be optimistic, but it's not out of the question if the roof has been well maintained over the years — especially if it's located in a benign climate like that in some of the drier western states. Recognize, too, that the modern materials you might replace this historic roof with would likely not be as durable as the original material. The more popular metals used today are coated carbon steel and aluminum. You can generally expect 40 to 60 years out of those if they are installed properly. However, no painted finish on those materials will last that long. Today's premium factory paint options will go 35 years at best.   — *Rob Haddock*

Attaching Metal Roofing

Q. When installing metal roof panels, should you screw or nail, and should the fasteners go through the flats or the ribs of the roof profile?

A. Our company definitely recommends screws over nails for attaching metal roofing to the purlins. Screws are much stronger than nails in terms of pull-out strength, they don't back out as readily as nails, and it's possible to get a better seal with screws than nails. Screws into the purlins or roof deck should always be installed in the flat of the panels. This may appear to be incorrect because it places the screw in the water-line. This isn't a problem, however, because the screws used for installing metal roofing have neoprene washers that compress to form a good seal. With some profiles, gasketed sheetmetal screws are installed along the rib at side seams between two panels. The sheetmetal screws lock the two panels together but do not penetrate into the purlins or deck below.

In general, you shouldn't put the screws through the ribs for several reasons. First, one of two things will probably happen when you try to compress the neoprene washer. Either the crest of the rib will dent or the washer won't compress properly. In either case, you will get a poor seal. Second, you will need long screws, which will cost more than short screws. Third, a long screw will be like a little lever because it is sticking up so high with its shank unsupported. As the metal roof expands and contracts, it is likely that, over time, the screw will snap off.   — *David Keener*

Leak-Free Metal Roofs

Q. I am installing steel roofing panels with exposed fasteners. Where it is necessary to install two panels between the ridge and the eaves, how much should the panels overlap?

A. A good rule of thumb for end-lapped panels is to use a 12-inch end lap. If the metal is going over purlins, or nailers, lay out the panel lengths so that the lap occurs over the support. The bottom panel should be long enough to allow for a 1-inch eaves overhang and extend about 6 to 7 inches past the up-slope edge of the end-lap support. The top panel should extend 2 inches past the down-slope edge of the end-lap support.

To prevent leaks, always install the fasteners 3 to $3^1/2$ inches up from the end of the overlapping panel, and apply a bead of butyl sealant just down-slope of the fasteners (between the weather exposure and the fasteners). I prefer pre-extruded butyl sealant tape, $3/32$ x 1 inch wide.

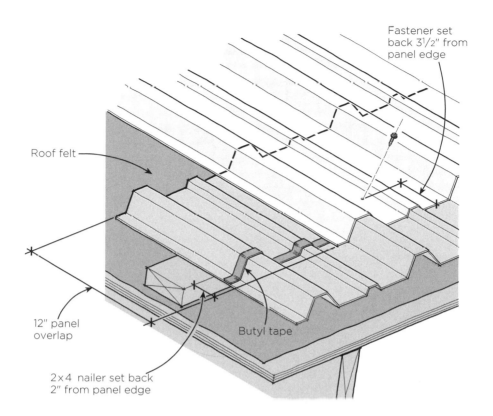

Fastener set back 3½" from panel edge

Roof felt

12" panel overlap

Butyl tape

2x4 nailer set back 2" from panel edge

This product is inexpensive, easy to apply, and doesn't hold the panels apart (causing fish-mouthing) at the panel lap. Butyl sealant doesn't cure and will remain tacky for the life of the roof. — *Daniel C. Jandzio*

Chimney Flashing for Metal Roofs

Q. I have to flash a chimney where it penetrates a corrugated metal roof with ridges 8 inches o.c. What's the solution?

A. The evenly spaced ridges of the roofing profile present a real challenge. When visualizing a flashing strategy for the down-slope face of the chimney, picture the plane of the roof at the top of the ridges, not at the lower, flat portion of the roofing profile. At the sides of the chimney, extend the flashing out past the nearest formed rib. After lining the top of the rib with butyl tape, fold the side flashing over the rib, then fasten through the flashing at the high point of the rib with neoprene-gasketed sheet-metal screws.

Fill the areas where the base flashing spans the flat portion of the roofing profile with the same closure strips that are used to seal off the

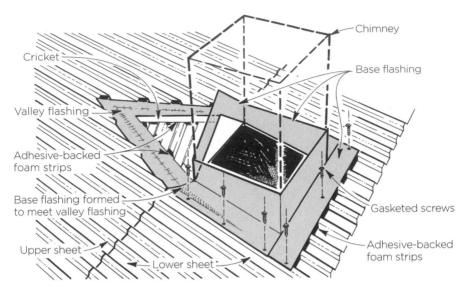

Chimney flashing over an agricultural metal roof has to follow the plane of the tops of the roofing ridges, not the flat parts. Fill the spaces between the flashing and the flat part of the roof with adhesive-backed foam strips. On the up-slope side, break the lower sheet of roofing even with the back of the chimney, and use a cricket and valley flashing to handle water from the upper roof.

metal ridge caps. These are adhesive-backed foam strips that conform to the profile of the metal roofing (see illustration).

When the chimney penetrates the roof below the ridge, use a cricket on the up-slope side to divert water around the chimney. The roofing, the cricket valley, and the chimney base flashing all come together at the chimney corners and must overlap in the proper order. First, fasten the lower sheet of roofing in place alongside the chimney, then install the base flashings at the bottom and sides of the chimney. As the side base flashing approaches the valley flashing, it will have to be shaped to lie in the lower, or "trough," section of the roofing so that the second sheet of roofing will lie flat. This results in an oddly shaped piece of base flashing, which I form using painted aluminum coil stock and a fuss-'n'-fit approach. It's not the most elegant piece of flashing, but it is leak free and suitable for the types of buildings where I install metal roofs.

Next, install the cricket valley flashing so that it overlaps the side base flashing. Finally, lay the upper sheet of roofing over the valley flashing. The roofing manufacturer can provide steel flat sheets that will match the color of the roof. However, these will most likely be tempered steel (like the roofing), and can be difficult to form, even when using a brake. On occasion, I've used painted aluminum coil stock for the flashing material. It's much easier to form and can be painted to match the roofing steel.

— *Carl Hagstrom*

Bending Metal Roofing to a Curve

Q. Can regular sheets of metal panel roofing be installed on a curved roof with a radius of 28 feet?

A. Regular metal roofing panels can often be installed in a curved application, depending upon the panel profile chosen and the radius of the curved roof. The minimum radius is 24 feet for through-fastened steel panels and 18 feet for through-fastened aluminum panels.

Standing-seam panels usually require a much larger radius — between 100 and 200 feet — for successful installation without pre-curving.

In general, panels with a shallower rib height and a continuous corrugation pattern are easier to curve than panels with distinct high ribs and flat areas. Panel ribs will tend to flatten out as they are bent over a curve, and care must be used to maintain equal width at both ends of the sheets.

Many curved roofs are actually arches, not uniform circular segments, and the calculated radius may not be accurate. If in doubt, try bending a scrap panel over the tightest radius section of the arch to see if it will work.

Finally, curved applications typically require additional fasteners to resist the forces induced by curving the panels. Increase the number of fasteners in the field panels to the fastening pattern recommended by the manufacturer for eaves and end laps. It may also be necessary to install stitch screws (short screws that join metal to metal without penetrating the substrate) in the side lap where separation is apparent. — *Daniel C. Jandzio*

Metal Roof Over Asphalt?

Q. I am bidding a job where the owner wants to install a new metal roof over a double layer of asphalt shingles. The roof has solid lumber rafters and plywood sheathing. Should I install 1x4 horizontal purlins over the asphalt and attach the metal roof to those? Should the purlins be pressure-treated?

A. To answer your first question: Purlins are always a good idea over asphalt shingles, for a couple of reasons. They allow you to create a flat roof plane and they provide solid wood to screw into where you need it. Also, asphalt shingles can be corrosive to metal roofs, so it's a good idea to separate the two materials.

Make sure you find a roof panel that is designed for use over purlins; some panels have to be installed directly over a plywood deck. Space the purlins according to the manufacturer's recommendations.

As for the size of the purlins, 2x4s are the best choice. Their 1 1/2-inch thickness gives the screws a greater pull-out resistance, making the roof much stronger against high winds. If you use 1x4s, you're getting only half as much screw penetration. Whatever purlin you use, it must be adequately attached to the roof deck and rafters.

It's probably not necessary to use pressure-treated purlins except in very hot, moist environments. But because most preservatives are corrosive to metal, if you do use pressure-treated wood, you must install a layer of roofing felt between the wood and the metal.   — *David Keener*

Stains on Metal Roofs

Q. A roof I installed has stains running down the steel standing-seam roofing below the skylights. What might be causing this, and how can it be prevented?

A. If the stains are rust trails, they could be coming either from the cut edges of the roofing or from the nails. More than likely it is from the nails, so use stainless-steel nails or screws to secure the flashing.

If the stains are dark, the corrosion is probably caused by a galvanic reaction between the skylight flashing and steel roofing panels. Some skylights come with preformed copper flashings that will react with the cut edges of the steel roofing panels in a relatively short amount of time. The stain could also be coming from a reaction between the nails and the flashing they are holding in place. To prevent problems, bend your flashings from the same type of coil stock as the roofing.   — *Henry Spies*

Flashing Tile Roofs

Q. What's the best way to flash a skylight on a mission tile roof?

Membrane and Metal

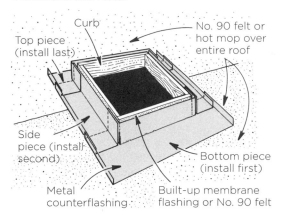

Curb

Top piece (install last)

No. 90 felt or hot mop over entire roof

Side piece (install second)

Bottom piece (install first)

Metal counterflashing

Built-up membrane flashing or No. 90 felt

Lead Flashing

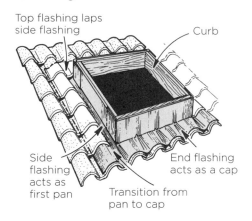

Top flashing laps side flashing

Curb

Side flashing acts as first pan

End flashing acts as a cap

Transition from pan to cap

Skylights can be flashed in two different ways. On roofs with a continuous membrane under the tiles, a simple metal flashing is sufficient to divert water away from the skylight opening (top). Without a continuous membrane, the curb must be flashed with lead (bottom). The side flashings act as the first tile "pan" and must transition to cap the tile near the bottom of the skylight.

A. Because of its fire resistance, tile is the predominant roofing here in southern California. On these roofs, we've done all our skylights using one of the two basic methods described here, both of which have performed well (no callbacks yet).

Membrane and metal. These days, most local building codes call for a continuous membrane over the entire roof, with the tile over the top. We commonly use either No. 90 felt or a hot-mop, three-ply built-up roofing for this water barrier. When we install a skylight, we start by framing in a 2x6 or 2x8 curb. The sides and ends of this curb get covered with the felt or the hot-mop tar. The felt or hot-mop then gets covered by formed metal flashing. The top, side, and bottom pieces of this flashing are shown in the illustration. Notice that water rolling down the sides and off the bottom of the metal flashing runs over the membrane and under the tile.

Lead flashing. If you don't want to get involved with the continuous membrane on the roof, then you'll have to opt for the expensive way to flash skylights. In this case, we install lead flashing that covers the curb and forms the first "pan" along the sides and top of the skylight. (Mission tile has both cap and pan tiles.) Near the bottom of the skylight, however, the lead has to transition from a pan to a cap, shedding water over the top of the row of tiles below the skylight. Ideally, you have a perfect, uncut cap tile starting on either side of the skylight over the lead flashing. But this method requires careful layout and a lot of time. *— Craig Savage*

Concrete Tile on Shallow Roofs

Q. We are getting a lot of jobs with Mansard-type roofs. The roof covering is concrete tile, but much of the roof has a slope of only 2:12. What precautions should we take?

A. According to the National Roofing Contractors Association, the minimum slope for a concrete, one-piece, barrel tile or interlocking flat-ribbed tile is 4:12. For flat shingle tile, it is 5:12. The *CABO One- and Two-Family Dwelling Code* prohibits the installation of any concrete tile on slopes less than 3:12, except with the permission of the building official. The best precaution is to use some other type of roofing on the low slopes. However, if a uniform appearance is desired, a double layer of No. 30 roofing felt, set in mastic or hot asphalt (or a continuous self-adhesive membrane) must be laid over the sheathing and under the tiles. This effectively forms a waterproofing membrane. The concrete tile would then be cosmetic and would protect the membrane from the sun.   *— Henry Spies*

Comparing Cedar Shingle Species

Q. How do eastern white cedar and western red cedar shingles compare for durability? I've heard that red cedar is more durable, but I wonder if the difference is enough to affect the service life of roofing or siding.

A. The heartwood (but not the sapwood) of both red and white cedar is naturally decay-resistant. Although the two species are listed in most handbooks as having heartwood with comparable rot resistance, the experience of many carpenters suggests that red cedar is a little more rot-resistant.

Several factors may account for this. When grading white cedar, it is quite difficult to distinguish between heartwood and sapwood, so it is likely that at least some sapwood slips into the all-heart grades.

Another possible factor affecting shingle durability is grain orientation. The best grade of red cedar shingles (#1) is all heartwood, clear, and vertical-grained. The best grade of white cedar shingles (extra clear) is all heartwood, clear, and typically flat sawn — although, like all flat-sawn lumber, some of the material is vertical-grained. Since wood shrinks and swells twice as much tangent to the growth rings as it does perpendicular to them, vertical-grain shingles lie flatter on a roof or wall after experiencing repeated wetting and drying cycles. That's why flat-grained white cedar shingles are more prone to cupping, splitting, and failure, and why I don't recommend using white cedar shingles on a roof.

I expect 30 years of service from a high-quality red cedar roof and much less from a white cedar roof. As siding, either red or white cedar shingles, if properly installed and maintained, should last a human lifetime. — *Paul Fisette*

Purlins vs. Plywood Under Shake Roofs

Q. For shake roofs, can I substitute purlins for plywood roof sheathing? Does the plywood serve as a diaphragm on a sloped roof?

A. In my view, 1x6 purlins are the best choice for a shake roof. Shakes last much longer when installed on purlins or skip sheathing because they can dry from the back as well as from the face. However, plywood is used more often these days because it is faster to install and it stiffens the roof structure.

Plywood roof sheathing on a sloped roof (particularly on low-slope roofs) provides some diaphragm action, increased significantly if the roof framing is blocked at the exterior wall line and continuous nailing secures the plywood to the blocking. However, because this is not always done, most light-frame structures are designed to be strong enough without relying on a roof diaphragm.

The IRC requires structural panel roof sheathing in seismic zones (Category D2) and in areas where the average winter temperature is 25°F or less (due to problems with ice and wind-driven snow). Truss roofs may also require solid sheathing as part of their permanent bracing. Where solid panel roof sheathing is required, the skip sheathing may be laid over 1x4 or 1x6 vertical nailers installed on top of the solid sheathing.   — *Henry Spies*

Underlayment for Shakes

Q. What underlayment should I use under cedar shakes?

A. A continuous underlayment is not recommended under shakes. Instead, 18-inch-wide strips of No. 15 or No. 30 felt are interwoven with the shakes. The bottom edge of the felt strips should start twice the exposure width above the shake butts (see illustration). For example, with 24-inch shakes having a 10-inch exposure, the "interlayment" should begin 20 inches above the butts. The felt interlayment should only be nailed to the sheathing along the top edge.

At the eaves, use a 36-inch-wide, No. 30, felt eaves flashing or an ice membrane. — *Henry Spies*

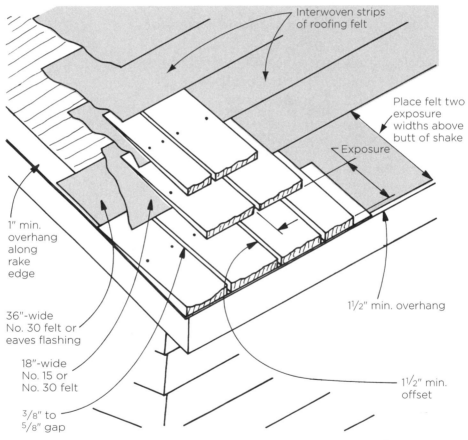

When installing wood shakes, interweave 18-inch-wide strips of roofing felt between each course. Nail the felt strips to the sheathing along the top edge, and keep each strip two exposure widths above the butt edge of the course below. This way, the felt isn't visible in the 1/2-inch gaps between the shakes.

Bending Wood Shingles

Q. I'm about to take on a wood-shingle roofing job on an unusual roof with a lot of sinuous curves. I'm planning on steaming and bending strapping to provide a nailing base for the shingles, then steaming and bending the shingles, too. But is there an easier way?

A. You can make slight bends in shingles by just soaking them. But for serious curves, you've got to steam them. Get a book on wooden boat building or visit a boatyard if you want to learn how to steam wood — the same techniques that work for big timbers work for shingles. I've seen people rig up a steaming apparatus from an old beer keg, a piece of waste pipe, and a propane heater — if you're handy, it's not too hard. Ten or 15 minutes of steaming will soften up shingles. Once they're soft and you've bent them into the shape you want, douse them with cold water and they'll hold that form.

However, when you start to get creative like this with shingles, you're departing from their original purpose as a water-shedding roofing material and emphasizing their visual effect instead. It's prudent not to push it. I've done a lot of decorative roof shingling, and in my experience, the best way is to start by providing a reliable waterproof layer under the shingles. In the old days, I would use two layers of No. 15 roofing felt with a layer of cold roofing mastic sandwiched between. Once products like Grace's Ice and Water Shield came out, I started using those instead (doubled up and lapped — I don't trust a single layer in freezing climates).

With a reliable waterproof substrate in place, you have a lot more leeway with how you trim and arrange your shingles. For instance, in places where you reduce your reveals and leave just a few inches exposed to the weather, there's no need for all that extra shingle that's buried under succeeding courses in a standard triple-coverage application. For sharp bends, I'll often cut quite a few inches off the thin end of the shingle to make it easier to work with. This way, I can avoid some of the steaming and soaking.

While cutting down your shingles can make this job a lot easier, it may compromise the water-shedding capabilities of the shingles somewhat (but that decision was already made by the designer who put all the curves in the roof plan). The shingles will still shed almost all the water, and they'll protect the membrane from the sun. But you'll be relying on the membrane as your final defense against the rain.   — *Martin Obando*

Wood Roofs on Dormers

Q. How should we detail wood shingles on dormers? Are the ridge shingles clipped on the same way as asphalt shingles? Is there a trick to getting the valley flashing to lie flat? What other details are important?

A. The ridge shingles should be attached with alternate overlaps, similar to the way a woven corner is detailed on a wall. The exposure should equal that of all regular courses of shingles. Note that longer nails will be necessary for the ridge shingles. Ridge shingles should not be clipped like asphalt-shingle ridge caps.

Valley flashing should overlap 6 inches at end joints, and flashing sheets should be no longer than 6 feet. To keep the metal valley flashings lying flat, do not nail them directly to the roof sheathing. Instead, fasten them with clips nailed to the sheathing and locked into the rolled edge of the flashing. This allows the flashing to move with temperature changes. The shingles should not be nailed through the metal.

The most critical part of installing a wood shingle roof, other than the flashing, is the nailing. Each shingle should have only two nails, each driven not more than 3/4 inch in from the edge of the shingles and no more than 1 inch above the exposure line. The nail heads should be driven to the surface of the shingle, not indented into the shingle fibers. The

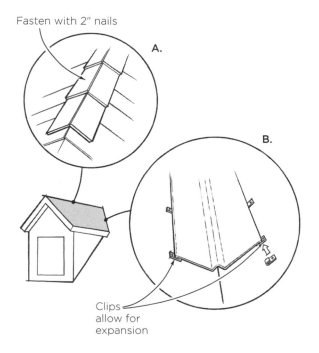

Fasten with 2" nails

A.

B.

Clips allow for expansion

Install wood ridge shingles with alternate overlaps (inset A). To accommodate expansion and contraction of metal valley flashing, fasten the metal to the sheathing with clips that are locked into the rolled edge of the flashing (inset B).

shingles should overhang the rake edge 1 inch and the fascia a minimum of $1^1/2$ inches. If gutters are used, the shingle may overhang by 1 inch. The slots between shingles should be $1/4$ inch to $3/8$ inch wide, and offset from the slot in the course below by at least $1^1/2$ inches. *— Henry Spies*

Turning Cedar Shingles Gray

Q. What's the best way to treat new white cedar shingles to give them a gray, weathered look?

A. If your customer is unwilling to wait for the natural weathering process, new cedar shingles can be either stained or treated with bleaching oil. Using a gray semitransparent stain is the fastest way to change the color of new shingles. It is important to choose a stain that includes a fungicide, since fungus can cause black stains on cedar shingles. Bleaching oil, which is available from Cabot Stains (800/877-8246, www.cabotstain.com), requires the action of sunlight to change shingle color. Bleaching oil accelerates the natural weathering process and will turn new cedar shingles a uniform gray in 6 to 12 months. *— Martin Obando*

Copper Corrosion on Cedar Roofs

Q. Can copper flashing be used with cedar shingles? I was told the natural oils in the cedar can corrode the copper.

A. The soluble tannins in cedar can cause corrosion of copper flashing, and the textbooks discourage using copper with cedar. However, I have seen dozens of roofs with a combination of cedar shingles and copper flashings where the flashing has lasted as long as the cedar shingles. In most cases, flow lines or patterns had been etched into the copper, but I have never found one that has corroded through. *— Henry Spies*

Moss on Cedar Shingles

Q. How can I kill moss growing on cedar shake roofs? And after I've killed the moss, what can I put on the roof to keep it from growing again?

A. Scrub the surface with a mixture of $1/3$ cup household detergent, 1 to 2 quarts of household bleach, and 2 to 3 quarts of warm water. That should kill the moss. The detergent helps wet the surface, and the bleach kills the growth. To prevent the moss from returning, you have to create conditions that won't support its growth. It's not possible to keep all dampness off the roof, but you can enable it to dry out faster by cutting back overhanging trees. You can also poison the food source of the moss by applying finishes that contain zinc, or a solution of copper naphthenate with 3% to 4% metal content. Unfortunately, those finishes don't last more than two years, so they require maintenance.

Another method is to install a ridge cap made of zinc or copper. "Bleed" water running down the roof surface as a result of the normal corrosion from those metals will help reduce moss growth. Tucking strips of copper under the butt end of the shingles every 4 to 6 feet down the roof, leaving $3/4$ to 1 inch exposed, may also help. — *Paul Fisette*

Life Expectancy of a Hot-Mopped Roof

Q. What is the life expectancy of hot-mopped asphalt on a flat roof? Will aluminum paint extend the life of the roof? If so, how often should the roof be recoated with the paint?

A. The average life expectancy of a hot-mopped asphalt roof is 15 years. If the roof drains well, its life span might be longer. The edge details are critical to the life of this type of roof. Joints in any edge metal must be maintained regularly because the metal expands and contracts at a different rate than the roofing, which tends to crack the roofing at the joints. An aluminum coating can extend the life of the roof because it reflects more heat, reducing alligatoring. The aluminum paint should be reapplied every three to five years. — *Henry Spies*

EPDM vs. Modified Bitumen Roofing

Q. Why is EPDM roofing so much more expensive than torch-applied modified bitumen? Is EPDM roofing that much better?

A. For a variety of marketing reasons, the bids you receive from local roofing contractors may not reflect national averages. Normally, the cost of a fully adhered EPDM roof should be fairly close to the cost of a modified bitumen roof. EPDM is a butyl-based rubber treated to withstand UV and direct exposure to the sun. An EPDM roof will probably outlast a modified bitumen roof or any other asphalt-based membrane. Asphalt products contain oil, which evaporates, eventually rendering the carrier dry and brittle. Frequent wet-dry cycles will accelerate that process, especially if the wet periods are prolonged. Other factors that can affect the longevity of a roofing membrane include the specific details of insulation, fastening, flashing, lap treatment, and maintenance. — *Harrison McCampbell*

EPDM Over OSB

Q. Does EPDM roofing always need to be installed over polyisocyanurate insulation, or can it be installed directly over OSB?

A. Although I prefer a base of polyisocyanurate, OSB or plywood will work fine under fully adhered EPDM. Install the OSB smooth side up. To ensure an even surface, I make sure the fasteners are flush, and install duct tape over the sheathing seams to soften any transitions that may occur if the sheathing begins to curl. There is no need to prime the OSB before installing the EPDM adhesive. EPDM sticks more tenaciously to OSB or plywood than it does to polyisocyanurate. If you are used to being able to adjust the EPDM slightly as it's installed, this can be a disadvantage: once the EPDM is cemented to OSB, you won't be able to pull it up for readjustment. — *Joseph Bublick*

EPDM Roofing Bubbles

Q. After installing an EPDM roof, I noticed that there are a few air bubbles under the roofing. Is this a problem? If so, is there any remedy?

A. In most cases, bubbles in EPDM roofing are just a cosmetic problem, as long as they aren't growing. Usually, bubbles are not associated with leaks.

The most common cause of bubbles is the application of rubber to the deck or insulation board too soon, before the adhesive is dry enough. In some cases, a contributing factor can be high pressure in the roof system, caused by wind uplift from above combined with air leakage from the building below to the underside of the roofing membrane. If this is the problem, the solution is to install pressure-relief vents. Pressure-relief vents, which are one-way air vents typically measuring about 2 to 3 inches in diameter, are available from roofing manufacturers.

To repair bubbles, cut them out and repair the areas according to the roofing manufacturer's instructions. — *Joseph Bublick*

Stripping an EPDM Roof

Q. I have to strip off an EPDM roof. Is there anything special I need to know?

A. No, there's nothing special about tearing off EPDM roofing. You use the same procedure you would use with any other roof. In fact, EPDM is probably the simplest roof covering to remove, because you can easily cut it with a knife and just pull the material up. Just like with any other roof system, though, there's no telling what you'll find when you look under it. If there's a substrate applied under the EPDM — typically, we apply a rigid insulation board with a tough protective skin — you'll want to strip that off, too, so you can inspect the structural deck. The substrate is usually screwed down, so you'll have to either back those screws out or cut them off.   — *Joseph Bublick*

Sidewall Flashing

Q. What is the best way to detail the connection of a pitched roof abutting a sidewall?

A. According to the National Roofing Contractors Association, when the rake edge of a pitched roof intersects a vertical wall, No. 15 felt should be laid under the shingles, extending 3 to 4 inches up the vertical wall. Metal step flashing (flashing shingles) should be used, with one flashing shingle for each row of shingles. The flashing shingle should be bent to extend under the asphalt shingles on the roof about 2 inches, and 4 inches up the vertical wall. Each metal shingle should be placed slightly up-roof from the bottom edge of the asphalt shingle that overlaps it. Nail the flashing shingle with one nail along the upper end of the vertical leg, as shown in the illustration.

The housewrap and the siding should be brought down over the flashing shingle to serve as a counterflashing, but cut the siding back a couple of inches, far enough to permit painting the ends of the wood siding. This will prevent water from wicking into the ends. If vinyl or metal

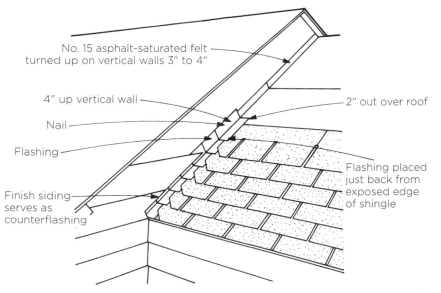

Step flashing should be used when a pitched roof intersects a vertical wall.

Chapter 3: Roofing/*Flashing*

siding is used, the flashing should be sealed to the sheathing with caulk or butyl tape. Vinyl and metal siding are not waterproof, and water blowing through the joints can run down the sheathing and behind the flashing (see "Weathertight Exteriors for Wind-Driven Rain," page 122).

— *Henry Spies*

Replacing Chimney Counterflashings

Q. Whenever I need to replace chimney counterflashings, I seal the top of the flashing with Quick Crete, using a caulk gun. This product is made for concrete repairs, and comes in a tube. I know that some people mix up a little mortar for this job, and others use silicone or urethane caulk. What is the best product to use?

A. Mixing up mortar is time-consuming and messy. Since the mortar can shrink as it cures, you don't know how watertight the joint will be. Products installed with a caulk gun are much easier to install. When replacing chimney flashings, I usually use a polyurethane caulk. Silicone caulk and products like Quick Crete also adhere well to old masonry. Some people prefer products made for masonry, because they have texture. Whether you choose a product like Quick Crete or a caulk, using a caulk gun allows you to get the product into the tight spaces between the old masonry and new flashing, providing a solid, watertight seal and a clean, finished appearance.

— *Tom Brewer*

Drooling Drip-Edge

Q. Most of the houses we build have 12:12 roofs with plumb fascias. Even though our roofer uses drip-edge, water still sometimes drools down the fascias. What are we doing wrong?

A. Standard drip-edge profiles include a turned-out bottom lip that directs water away from the fascia. The distance from the end of this lip to the face of the fascia varies, depending on the manufacturer and the style of the drip-edge you are using, from as little as 1/4 inch to as much as 1 inch. Most drip-edge profiles are successful at directing water away from the fascia during periods of heavy rain. During drizzly weather, though, rainwater is sometimes able to curl under a drip-edge and run

Rite Flow drip-edge

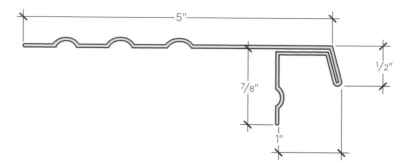

down the fascia, especially if the drip-edge has a short lip. One manufacturer selling a drip-edge with an aggressive lip is Lamb & Ritchie of Saugus, Mass. (781/941-2700, www.lambritchie.com). Its Positive "Rite Flow" drip-edge has a lip that projects 1 inch from the fascia. It's available in 30-gauge galvanized, white galvanized, and brown galvanized steel; 26-gauge mill-finish; white and brown aluminum; and 16-ounce copper. — *Harrison McCampbell*

Connecting Gutter Downspouts to Footing Drains

Q. Can gutter downspouts be connected to 4-inch pipes connected to the footing drains? The footing drains eventually terminate at daylight, away from the house.

A. This is a terrible idea. Don't do it. Downspouts should feed into a solid, nonperforated pipe that directs roof water far away from the house. The pipe should run independently from the footing drain.

Surprisingly, there is nothing in most building codes that prevents you from connecting downspouts to a foundation's perimeter drainage system. The pipes used for perimeter drainage have slits or weep holes along the pipe to allow water to enter the pipe and be carried away. The goal is to remove water from around the foundation. Bringing additional water from connected downspouts to the base of the foundation runs counter to this goal, since it introduces water to a place where you don't want it. In all likelihood, the downspout water would leak from the perforated pipe, saturating the soil near the footing. If the perimeter footing drain ever fails or gets blocked, you also run the risk of flooding the subsoil area.   — *Paul Fisette*

Gutters in Cold Climes

Q. The jury still seems to be out as to whether to install gutters in a northern climate. Could you please shed some light on this, especially as to how omitting gutters would contribute to leaky basements?

A. Gutters are helpful and are required in many jurisdictions, but they can cause serious problems in very cold climates. They can fill with ice and cause water to back up under the roof sheathing even if a membrane like Grace Ice & Water Shield has been installed under the shingles. They can be knocked down by ice and require yearly repairs. The downspouts are often cracked open and discharge water onto the siding, which can lead to rot.

I typically recommend gutters only for houses that are very well insulated, have effective attic ventilation, and have gone through one or more winters without ice damming.

Removal or omission of gutters does not have to create basement water problems. Even if you have gutters, you can have a damp or leaky basement if the grading, walks, driveways, decks, or planters allow water to stand against or run toward the foundation.

There are four main criteria for safely eliminating gutters. First, make sure the grade slopes away from the building at a rate of 2 inches per horizontal foot for as far as is practical. When building new, set houses higher to start with, so you can grade to them. Second, plant a healthy stand of grass or ground cover on the sloping grade for a few feet, starting at the foundation. This is preferable to planting bushes and flowers in flat mulched beds, which allow water to stand and percolate deep. Instead, plant flowers and shrubs a few feet away from the foundation; that way, the homeowners will be able to see and enjoy them from inside, as well. Third, set flagstones or other paving material in the sloping grade at the drip line of the roof to catch the brunt of the roof water and deflect it. (However, if the roof overhangs are too narrow, this approach can cause splashback and deterioration of siding and wood trim.) And, fourth, make sure that the foundation wall is protected with dampproofing or waterproofing, along with footing drains. These measures should ensure that the basement remains dry, unless there is an underground spring or high seasonal water table.   — *Henri de Marne*

Ice-Proof Gutters

Q. How should gutters be attached to prevent them from getting ripped off the fascia when snow slides off the roof during winter?

A. Install gutter hangers at least 24 inches on-center. Make sure the spikes fasten through the fascia to a 2x subfascia or to rafter or truss tails. Also, mount the gutters so that a straightedge laid on the roof extends above the outer edge of the gutter (see illustration). This will help prevent snow slides from taking the gutter off. A well-insulated and well-ventilated "cold roof" will have fewer gutter problems, as the snow on such roofs tends to melt gradually and not refreeze to cause ice dams.   — *Henry Spies*

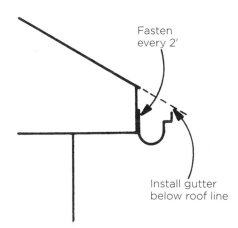

Fasten every 2'

Install gutter below roof line

To prevent snow slides from tearing the gutter off, install gutters below the roofline, as shown.

Roof Ventilation

Roof Vent Requirements

Q. By code, how many linear feet of ridge vent are needed per square of roofing?

A. The amount of ridge vent needed is not related to the roof area; it is related to the living space below the roof. Most codes (including the International Residential Code) require roof ventilation that has a minimum total net free ventilating area of $1/150$ of the ceiling area of the living space beneath it. This can be reduced to $1/300$ of the area if 50% to 80% of the required ventilation area is provided by ventilators located in the upper portion of the space to be ventilated, at least 3 feet above the eaves, with the balance of the ventilation provided by eaves or cornice vents. As an alternative, the net free cross-ventilation area can be reduced to $1/300$ when a vapor barrier having a transmission rate not exceeding 1 perm is installed on the warm side of the ceiling.

The net free ventilation area of a ridge vent is determined by the manufacturer. Wind-tunnel tests indicate the effective net free area of ventilation provided by most ridge vents is typically half that indicated by the manufacturer. As long as the ceiling air/vapor barrier is well sealed, there is no problem with overventing the attic area, so the best course is to provide as much ridge vent as can be installed on the roof. *— Henry Spies*

Eaves Vents With No Soffits

Q. I'm reroofing and residing a small home with no soffits. The customer wants to add attic ventilation. I have no problem adding the ridge vents, but there is no soffit overhang to put the eaves vents in. In fact, to have enough space for the vinyl J-channel, I'm having to add an extra 2x4 fascia. How can I get airflow into the attic?

A. If you're reroofing anyway, you might try Air Vent's vented drip-edge (800/247-8368, www.airvent.com) or Lomanco's Starter Vent (800/643-5596, www.lomanco.com). This product extends farther than standard drip-edge and has vent intakes in the underside of its horizontal projection. If you drop the fascia an inch or so below the top of the rafter plumb cuts, you may be able to sneak some air into the attic.

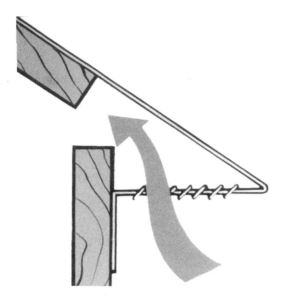

Lomanco's Starter Vent is a combination drip-edge and continuous soffit vent.

The best solution, if possible, is to extend the roof overhang, which will improve ventilation and help protect the siding, windows, and trim from excessive water exposure.

— *Don Jackson*

Can Wind Short-Circuit a Ridge Vent?

Q. Wouldn't a single-sided ridge vent be more effective than a standard ridge vent for a home built on a windy site? It seems that with openings on both sides of the ridge, a standard vent would simply short-circuit when the wind blows rather than draw air from the eaves vents.

A. In order for a ridge vent to exhaust, you need a pathway and a reliable driving force. Both a single-sided and a double-sided ridge vent provide a pathway, so the important question is this: What is the main driving force that pushes attic air up and out of the ridge vent? While wind direction can induce roof venting — and it's been shown that even soffit-only venting can draw air out of roofs — most often the answer is the buoyancy of the attic air.

If there is a fair amount of heat loss from the house into the attic, then the buoyancy of the warm air rising causes it to escape at the highest point, the ridge. However, a more energy-efficient house experiences less heat loss, so in that case buoyancy becomes less of a driving force.

Based on some tests I've run, I think it's important to install ridge vents that have an external baffle, such as ShingleVent II (Air Vent Inc., 800/247-8368, www.airvent.com). As wind passes over the roof ridge, the

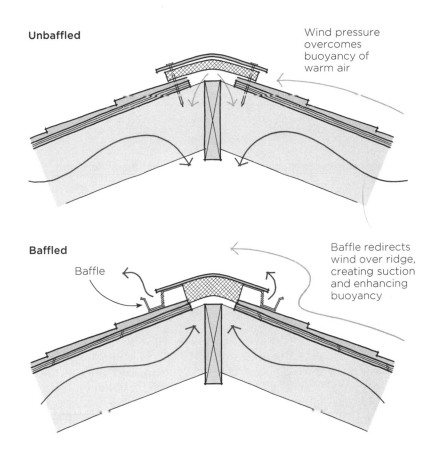

Unbaffled

Wind pressure
overcomes
buoyancy of
warm air

Baffled

Baffle

Baffle redirects
wind over ridge,
creating suction
and enhancing
buoyancy

airstream jumps over the vent's baffle, causing suction as the air lifts
upward — the same principle involved in how an airplane wing works (see
illustration). Called the Bernoulli Effect, this driving exhaust force works
regardless of wind direction. Without an external baffle, either a single- or a
double-sided roof vent can allow outside air to come in and short-circuit
the venting process. — *Paul Fisette*

Can You Combine Ridge and Gable Vents?

Q. I'm reroofing an older ranch house that has
continuous soffit vent panels and gable-end attic vents
but no ridge vents. I'm wondering if I should add a
continuous ridge vent; it would be fairly easy to do
while the roof is stripped. Is it okay to combine soffit-
to-ridge roof ventilation with gable-end vents?

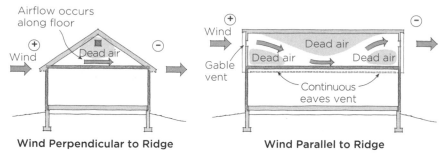

Wind Perpendicular to Ridge   Wind Parallel to Ridge

On a house with eaves and gable vents but no ridge vents, wind that is perpendicular to the ridge tends to create airflow along the floor of the attic, but leaves hot dead air in the roof peak (left). When the wind is perpendicular to the gable, the cooler outside air enters the gable vent and drops to the floor of the attic before rising to exit at the other end, again leaving zones of dead air (right).

A. Upgrading existing ventilation when reroofing is a smart move — an opportunity that is missed all too frequently. But should you add a ridge vent to a roof with gable vents? No way — that's a prescription for disaster. Gable vents will alter the airflow around the ridge vent and, especially when wind is parallel to the ridge (at right angles to the gable), can actually reverse airflow through the ridge vent, pulling rain or snow into the attic. Unfortunately, however, even when they're matched with continuous eaves venting, gable vents are not very effective because most of the airflow is along the floor of the attic. This leaves much of the attic volume unvented (see illustration), with pockets of dead air that can store summer heat and radiate it into the living space below. The most efficient option is ridge venting combined with continuous soffit vents. If, in this case, you choose to add ridge vents, you must either remove the gable vents or seal them up from inside the attic. Sealing and leaving them may look better from outside, to avoid creating a blank gable. There are also some attractive, strictly decorative, gable vents available.   — *Mike Keogh*

Wide Ridge Vent

Q. I have a job involving a ridge beam built up out of four pieces of LVL. Is there a ridge vent available that will span the wide opening this requires in the roof?

A. You could use a system I devised years ago before the advent of variable pitch ridge vents, a design which Air Vent Inc. (800/247-8368, www.airvent.com) borrowed and illustrates in their product literature for using utility vent on steep roofs.

Rafters Above LVL

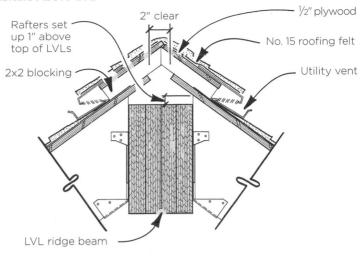

Rafters set up 1" above top of LVLs

2x2 blocking

2" clear

½" plywood

No. 15 roofing felt

Utility vent

LVL ridge beam

Rafters Flush With LVL

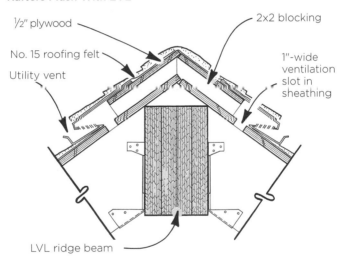

½" plywood

No. 15 roofing felt

Utility vent

2x2 blocking

1"-wide ventilation slot in sheathing

LVL ridge beam

You have the choice of setting the top of the rafters up 1 inch above the top of the LVL so as to allow an air space, or of leaving the rafter tops flush with the top of the LVL and simply carrying the plywood to the ridge, where the two sides will butt (for extra support you can add triangular blocks on top of the ridge beam). If you choose the second option, cut a 1-inch-wide ventilation slot at the appropriate spot on each side of the ridge (see illustrations, above).

The ridge vent is composed of sections of Air Vent's Utility Vent (it comes in 8-foot lengths, four to a package, with connector pieces and filter material) installed on each side of a site-built mini-roof. We found it easiest to build this mini-roof in 8-foot sections on the ground, attaching 1/2-inch plywood to 2x2s, 24 inches on-center. After nailing the mini-roof to the ridge, we installed the vent strips over the plywood, and covered the whole thing with No. 15 felt and roof shingles. We used this system on my own house in 1979 and on a number of other jobs before adjustable-pitch vents became common. It is still one of my favorites because it is externally baffled and is as strong as the roof itself.   — *Henri de Marne*

Venting Hip Roofs

Q. Can I use a ridge vent to vent a hip roof?

A. The short answer is yes. I know that venting hip roofs adequately can be difficult, and that many people think that the relatively unobtrusive ridge vents are a better aesthetic option than a bunch of unsightly mushroom vents.

However, I'm not a fan of using ridge vents along hips. I installed ridge vents on the hips of two homes 11 years ago. About six months after the homes were complete, both homeowners had water stains on the second-floor ceilings. I found the fiberglass insulation damp beneath a couple of the hips, and I could see where water had dripped along the hip rafters. I thought the leaks were due to a particularly severe thunderstorm and figured it wouldn't happen again. But two weeks later, during a moderate storm, the leaks recurred. I removed the hip vents and haven't had a problem since.

My present roof-venting strategy is to use a continuous soffit vent (either a strip vent or a fully vented vinyl soffit) and a ridge vent on all true ridges. (I haven't had any callbacks from leaking ridge vents installed on actual ridges.) Some ridge vents work better than others. I think ShingleVent II by Air Vent Inc. (800/247-8368, www.airvent.com) is better than many of the others. I've found some of the roll-type vents subject to installion errors that reduce the net free vent area.

On a hip roof that lacks enough of a ridge for adequate venting, I install roof vents (mushroom vents) cut high on the roof slopes that aren't likely to be viewed.

One of the reasons for installing attic venting is to remove moisture vapor that can condense on roof framing, potentially causing rot in cold climates. If you stop moisture from getting into the attic in the first place by tightening up the ceiling, then you can reduce the need for attic venting. Moisture "piggybacks" on the air that leaks into the attic. The biggest air leaks you'll find are at attic access panels or pull-down stairs, standard recessed light cans, regular light fixture electric boxes, HVAC ducts and chases, whole-house fans, and the spaces around pipes and wires that run through the top plates of walls. During new construction, these air leaks are easy to address.   — *Mike Guertin*

Ice Buildup Problem

Q. We built a custom home for a client in west Michigan a couple of years ago, and the home has had problems with ice ever since. It's a 1,450-square-foot ranch with cathedral ceilings and many can lights throughout. We used blown fiberglass insulation in the ceiling assemblies. From the beginning, the can lights (IC-rated) overheated and tripped their thermal-protection breakers. We finally resorted to pulling the insulation away from the housing of the can lights so they will not overheat. But as a result, the heat from these lights now warms up the roof and has created a horrible ice problem instead. Last fall we even added four pot vents to the back of the roof in addition to the soffit-to-ridge venting. The pot vents have improved the situation but not completely. I drive by this home frequently and see ice buildup there while other homes in the area are ice-free. What can we do?

A. The fact that the home your company built is the only one in the neighborhood with ice problems is not going to help your reputation as a quality builder. Adding pot vents to the roof is tantamount to heating the outdoors in an effort to cool the can lights. Since you seem to be able to reach the can lights from an adjoining attic area, you might be able to place a sealed, insulated box around each can, but there's a chance the overheating problem would come back. Instead, line up your electrician and drywall finisher. Pull the cans and convert them to track lighting or some other kind of surface-mounted fixtures, then seal up and patch the ceiling, making sure you plug all the holes. Carefully replace the insulation and get rid of the pots, and your problem should disappear. In the future, avoid can lights in cathedral ceilings. — *Don Jackson*

Exteriors

4

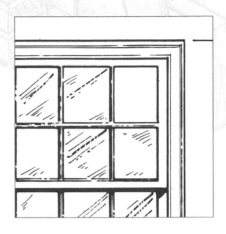

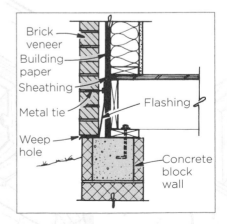

Brick veneer
Building paper
Sheathing
Metal tie
Weep hole
Flashing
Concrete block wall

Weathertight Exteriors for Wind-Driven Rain

Q. As a general contractor in the Blue Ridge Mountains of southwestern Virginia, we often build vacation homes at altitudes higher than 4,000 feet where wind-driven rain is a regular weather feature. On one site, the wind regularly reaches 50 to 80-plus miles per hour and actually blows rain uphill. We have a south-facing window wall full of fixed-glass and awning windows. The wall has 2x6 studs, 1/2-inch OSB sheathing, Tyvek housewrap, and 1x10 horizontal shiplapped pine. With only a 3/8-inch lap on the siding, I can imagine that water might be driven behind it. But how is it getting past the Tyvek and OSB through the wall? Water drips from the interior window head jambs, and with the interior wall paneling removed, it can be seen on top of the sole plate. We had to replace some buckled hardwood flooring after a vicious storm last January, and before we repair it again, we want to make sure the wall won't leak. Are there any methods or materials that you could recommend?

A. We build homes along the Delaware coast, where storms bring heavy rains and winds of 40 to 80 mph on a routine basis plus the occasional hurricane. We take a few extra precautions to keep our houses dry:

- Back-caulk the window flanges at the top and sides as you install the windows. Do not caulk along the bottom flange, as this can trap water that might leak through the window unit itself. (Note: caulk is not an air sealant; use foam between the window and the rough opening to air-seal the opening.) Cut a slice in the wrap just above the window and install a piece of tar paper or metal flashing that goes behind the wrap and in front of the flange (see Illustration A). If there is trim above the window, extend the flashing over the trim, and tape the slice in the wrap. For best protection, use a sill pan below each window, extending the pan over the housewrap below.
- Under shingle siding, on the sides that get the worst weather (east and northeast for us), one option is to install tar paper over the housewrap for an extra layer of protection. Sidewall shingles let more water past

than bevel siding. Currently we use Benjamin Obdyke's Home Slicker to provide a vent space and drainage plane behind all wood siding that isn't back-primed (800/346-7655, www.benjaminobdyke.com).

- Watch out for step flashings, since they usually go on after the housewrap. It's important to slice the housewrap and install the step flashings behind it. In tricky areas where valleys channel water up against sidewalls, we lift the housewrap out of the way and install a backup strip of Grace's Ice and Water Shield over the step flashings and behind the housewrap (Illustration B).
- Caulk between the siding and the window trim along the side casings, but never along horizontal joints at the window's top and bottom. Hope for a bad storm before the drywall goes up to give your weatherproofing a test run.

— Patricia McDaniel

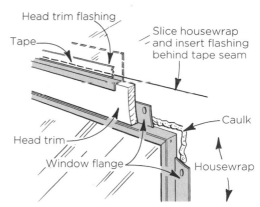

Illustration A. *To prevent leaks from wind-driven rain, use only windows with an integral nailing flange, caulking the flange to the housewrap. Above the window, slice the housewrap and slide the head flashing behind, and then tape the seam.*

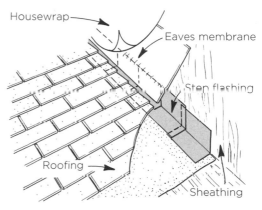

Illustration B. *To protect dormer sidewalls, make sure the step flashings are installed behind the housewrap. At areas where water accumulates, lift the housewrap and add a strip of self-adhering eaves membrane over the step flashing.*

Housewrap Behind Vinyl Siding

Q. Is it necessary to put housewrap behind vinyl siding on new construction?

A. Many builders and material suppliers think that using housewrap under vinyl siding is a waste of money. I disagree. For starters, a building wrap is required by code. If this provision has not been adopted in your jurisdiction yet, it will be soon. Second, it makes good sense. Housewrap is cheap insurance for two reasons: it protects the building shell from the weather and it helps save energy.

Vinyl siding is a good siding material, but it is loose-fitting and somewhat permeable to weather. Wind-driven rain can penetrate at overlapping ends, around windows at J-channels, at corner posts, at the intersections between siding and rake trim, and at any other places where joints are involved. A smart installer should expect water to penetrate the siding at times; the housewrap provides a backup weather barrier that allows water to run down and out without penetrating the structure. Proper flashings should be used at doors, windows, and other penetrations, and the housewrap should overlap the tops of these flashings.

The second issue relates to energy performance. Studies have shown that air leakage is an important factor in heating and cooling losses; installing taped housewrap directly over wall sheathing is a good way to modestly reduce air leakage. If your budget allows, a layer of taped foam insulation, instead of housewrap, will not only reduce air leakage but also improve the conductive resistance of the wall. — *Paul Fisette*

Asphalt Felt Under Vinyl Siding

Q. Can asphalt felt be used directly under vinyl siding? I have heard that the two products are incompatible.

A. The installation of asphalt felt as a weather barrier under vinyl siding has been common for years and is approved by the Vinyl Siding Institute's installation guide. I have never heard of any compatibility problems between the two products. During my years as an installer, going back to the late 1970s, I installed vinyl siding over asphalt felt on at least 400 houses and have never heard of any problems arising from the use of felt under vinyl. — *Bob Werner*

Vinyl Siding Over Fan-Fold Foam

Q. Before installing vinyl siding over existing wood siding, we first install a layer of fan-fold foam insulation. Is the fan-fold foam an adequate weather-resistive barrier?

A. Rigid foam will shed any water that gets past leaky vinyl siding. To be an effective weather-resistive barrier, though, the seams of the fan-fold must be carefully taped, since any seam can provide a pathway for water

to reach the structure under the foam. As long as the foam is not foil-faced, the best tape is contractor's tape from 3M. It's always a good idea to check with the foam manufacturer to be sure that the tape you use is compatible with its product. Fan-fold is typically sold in 1/4-inch by 4-foot by 50-foot sections and is installed horizontally across the existing siding to level out irregularities before residing with vinyl. This creates a horizontal seam every 4 feet. I have not seen any studies that show how well tape adheres to foam over the long haul. I am mildly concerned that if the tape fails and a horizontal joint is not flush (with the lower panel sticking out beyond the upper panel), a seam could trap water. To avoid this potential problem, you may want to consider installing the fan-fold vertically. All penetrations, as well as areas around doors and windows, require careful detailing. If the fan-fold and J-channel are simply butted against existing window and door casings, water can get between the J-channel and the casing, and then behind the fan-fold. Proper flashing details at windows and doors may require installation of new flexible flashing, head flashing, and casing.
— *Paul Fisette*

Two Layers of Housewrap

Q. I plan to install new cedar shingle siding over existing T&G board siding. Should I install housewrap under the new shingles? Since there is already a layer of housewrap between the plywood sheathing and the board siding, I'm worried that two layers of housewrap may lead to moisture problems.

A. I would recommend that you use a layer of No. 15 asphalt felt to cover the T&G boards. The new siding will not keep all rainwater out, especially in a heavy soaker of a storm, and the T&G boards need to be protected from water that penetrates the siding. If you installed a second layer of plastic housewrap instead of asphalt felt and liquid water were to leak into the region between the two layers of housewrap, the T&G boards could have a difficult time drying out. Asphalt felt is forgiving. Initially it blocks vapor and liquid water, but if it does get wet (under extreme exposures) it first stores water, then allows liquid and vapor to pass, and ultimately dries out. The perm rating of asphalt felt is about 5 (fairly low) when it is dry, but it rises to 60 (fairly high) as the felt nears saturation. Fussy installation and detailing is important. Carefully lap the asphalt felt so it sheds water, and be sure the flashing at the window heads is tucked under the felt.
— *Paul Fisette*

Wall Flashing for Exterior Chimneys

Q. What's the right way to flash an exterior masonry chimney to the wall of the house?

A. I bend an F-shaped piece of flashing that catches the ends of the bricks where they meet the house (see illustration). I use copper, lead-coated copper, or aluminum and bend the flashing in long pieces on an 8-foot brake. We nail the flashing to the sheathing over the housewrap, and then use a flashing tape to seal the outside leg of the flashing to the wrap. The mason spaces the brick away from the flashing so that any water that gets in can drain down and out the bottom through metal weep channels embedded in the mortar under the first course of brick. Depending on the height of the house, it may take three or more lengths of flashing to reach the roofline. At the joints, I prefer to butt the pieces rather than overlap them, because it looks better. To maintain a tight seam on the exposed kick where the flashing meets the brick, I run a spline several inches long inside the bend to span the joint.   — *Fred Seifert, Sr.*

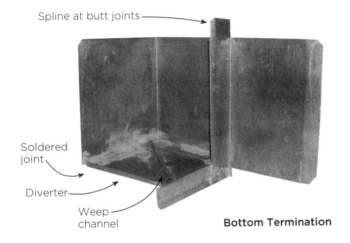

Spline at butt joints

Soldered joint

Diverter

Weep channel

Bottom Termination

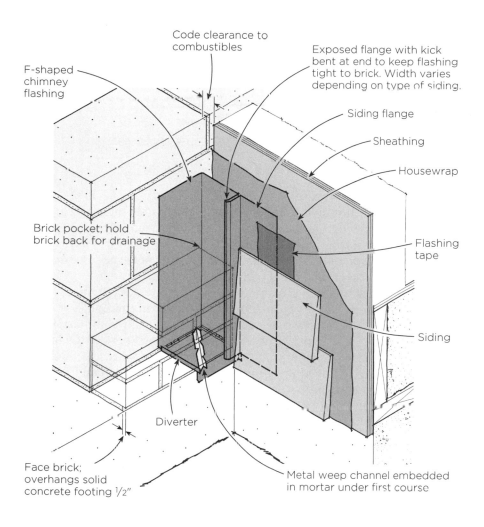

Code clearance to combustibles

Exposed flange with kick bent at end to keep flashing tight to brick. Width varies depending on type of siding.

F-shaped chimney flashing

Siding flange

Sheathing

Housewrap

Brick pocket; hold brick back for drainage

Flashing tape

Siding

Diverter

Face brick; overhangs solid concrete footing 1/2"

Metal weep channel embedded in mortar under first course

Vinyl Siding Over Foil-Faced Rigid Foam

Q. Can vinyl siding be installed directly over foil-faced rigid foam?

A. Some vinyl siding manufacturers void their warranty for heat distortion when their siding is installed over foil-faced foam. One study has shown that when vinyl siding is installed over foil-faced foam, the temperature of the back of the siding can be 8°F warmer than when it's installed over foam without foil. Installers who are concerned about heat buildup should select rigid foam without foil facing.

Nevertheless, some manufacturers have no objection to the installation of vinyl siding over foil-faced foam. Whenever vinyl siding is installed over rigid foam (with or without foil facing), studs should be spaced no more than 16 inches on-center and the thickness of the foam should be limited to 1 inch. Because rigid foam doesn't provide much resistance to vinyl siding sagging under its own weight, such an installation requires threaded nails that are long enough to pass through the rigid foam and penetrate at least 1 inch into the sheathing and studs.

Newer building codes require a secondary weather barrier (asphalt felt or housewrap) under vinyl siding, so check with your local building authorities before proceeding. Where allowed by your inspector, rigid foam will perform adequately as a weather-resistive barrier as long as all seams and flashing are taped with foil tape. — *Paul Fisette*

Shingle Siding Layout

Q. How do you lay out shingle siding so the courses break evenly above and below windows? And what's the best way to secure the row of shingles below a window so the nails aren't exposed?

A. Follow this procedure to lay out shingle siding so the courses break on the same line as the window trim: First, measure the window height and add the width of any trim you might have, then divide by the shingle reveal. Most of the time, you won't have an even number of courses, and you'll have to adjust the shingle reveal slightly. To do that, ignore the fraction and choose the closest whole number; this will be the number of even courses that will fit between the window trim. Then pull your tape

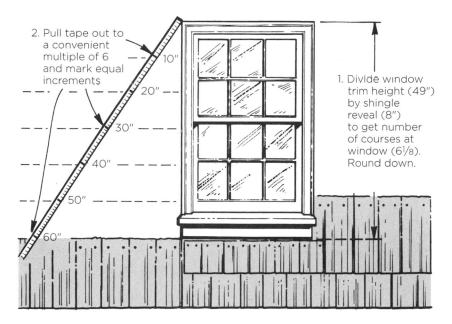

2. Pull tape out to a convenient multiple of 6 and mark equal increments

10"

20"

30"

40"

50"

60"

1. Divide window trim height (49") by shingle reveal (8") to get number of courses at window (6⅛). Round down.

Illustration A. Even shingle layout: *To find the number of shingle courses between the top and the bottom of a window: (1) Measure the window height plus the trim. Divide this number by the average shingle reveal. If the resulting number is a fraction, choose the nearest whole number. (2) Hold a tape measure at the corner where the head trim will be, and pull it out to an even multiple of six. To find the course lines, mark off the even multiples.*

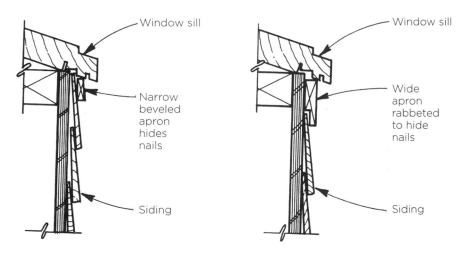

Window sill

Narrow beveled apron hides nails

Siding

Window sill

Wide apron rabbeted to hide nails

Siding

Illustration B. Apron trim: *To cover the nails holding the first course of shingles below a window, the shingle must be overlaid with trim. For narrow trim, rip it at an angle, so its surface stays flush with the side trim (left). For wide trim, cut a rabbet to conceal the nail line (right).*

measure out to an even multiple of that number of courses, run it at a diagonal next to your window (Illustration A), and tick off each multiple. These marks will be your course lines.

Let's take an example: Say your window height, including the head trim and apron, is 49 inches and your average shingle reveal is 8 inches. Since $49 \div 8 = 6^1/8$, you will have six courses (ignoring the fraction). To find the exact width of each reveal, pull the tape out to some large multiple of six, say 60 inches, and run it at a diagonal from the point that will be the corner of the window head trim to a level line extended from the bottom of the soon-to-be-installed window apron. Mark the sheathing at 10, 20, 30, 40, 50, and 60 inches. You've now got even course lines that will break on the window trim.

Repeat this procedure for the space between first- and second-floor windows, for the space between the trim for the second-floor windows, etc. Then, so you don't have to tick off all these different numbers all over the house, transfer the tick marks to a couple of long pieces of lumber, making two story poles. You can then leapfrog these as you work your way around the building.

As for concealing the nails, the only way I've found to do this is with a piece of trim. Since the trim on the sides and at the top of the window won't necessarily lay over the siding, but the apron trim will, I usually rip the apron at an angle (if it's narrow) or notch it (if it's wide), as shown in Illustration B. *— Sal Alfano*

Nailing Cedar Shingles

Q. A new lead carpenter on my crew says that when he learned to install cedar shingle siding on a house, he was taught to always put a third nail in the wider shingles. Is he right?

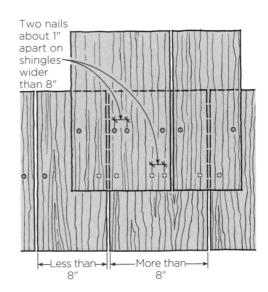

Two nails about 1" apart on shingles wider than 8"

←Less than→ ←More than→
   8"   8"

A. A shingle less than 8 inches wide should receive only two nails. However, a shingle 8 inches or wider should receive four nails. The extra two nails should be placed near the center of the shingle, about an inch apart. Experience has shown that two nails in the center of a wide shingle are less likely to cause the shingle to split than a single nail in the center. The Cedar Shake & Shingle Bureau recommends the four-nail approach to shingles wider than 10 inches.

— *Martin Obando*

Caulking Siding — or Not?

Q. When installing cedar clapboards, should gaps between the siding and trim be caulked?

A. Clapboards should be installed tight against the trim boards, and the joints should be left uncaulked. Omitting the caulk allows any water that penetrates the joint to drain reasonably well and promotes drying.

Although the use of caulk does not cause any disasters, it can elevate the moisture level of the sheathing. I have taken moisture readings of sheathing on many homes, and on average the sheathing underneath caulked corners has a higher moisture content than that under uncaulked corners, and a higher moisture content than the sheathing a couple of feet away from the corner. I recommend that vulnerable areas of the sheathing (like corners) should have a double layer of asphalt felt or housewrap under the trim boards.

Wide temperature and humidity swings and wet/dry cycling take a toll on caulked wood joints. In most cases, caulk quickly loses its bond from one of the wood surfaces and leaves a crack where water can enter. Yet even cracked caulk can slow the drying process.

Well-detailed caulk joints require several fussy details. For caulk to work well, the end grain of the clapboards should be painted in order to prevent absorption of the caulk solvent. Moreover, a bond breaker (backer rod) should be installed in the gap against the sheathing to avoid three-sided adhesion. This backer rod ensures that the caulk will adhere only to the siding and the trim board, so that the caulk can stretch without tearing.

A caulked joint works best when the gap measures a uniform 1/4 inch. But technical specs for most building sealants indicate that joint movement should not exceed 50% of the joint width. Since a 6-inch trim board can shrink and swell by 1/4 inch as a result of normal exposure to the elements, you would have to leave a 1/2-inch gap between the trim board and clapboard ends to accommodate 50% movement. Who's going to do that? Finally, uncaulked joints require less maintenance. Few homeowners have the time to dig out old cracked caulking on a regular basis.

— *Paul Fisette*

Silicone Caulk on Cedar

Q. I have heard that silicone caulk should not be used with cedar trim. Is this true? If so, why?

A. Silicone caulk is not the best choice to use with any wood trim. It works best in sealing the joint between two pieces of nonporous material, such as ceramic tile, metal, or glass. The oil in cedar will reduce the adhesion of silicone caulk, which should be applied to clean, dry surfaces. An acrylic or urethane-based caulk works best where one of the materials being sealed is wood.

How the caulk is applied is even more critical than the type of caulk used. The caulk should be applied so it bonds to two faces across a gap. That way, movement of the materials will place the bead in tension or compression, rather than peel the caulk away from one of the surfaces. A three-sided fillet of caulk in a corner will almost always fail. Best practice calls for detailing siding to avoid caulk altogether.   — Henry Spies

Applying Siding on a Curved Wall

Q. I am trying to put 1/2x4-inch redwood beveled siding on a wall with a 4-foot 8-inch radius. It doesn't seem like a very tight radius, but so far I have put up only three courses before a board splits out. It seems as if each piece wants to pull further from the wall at the top. I have some pieces soaking in a tub of water, hoping that will make them more pliable. Do you have any suggestions?

A. Beveled siding will not bend properly around your curved walls because of the gap created by lapping each piece over the piece below. You are, in effect, increasing the circumference of the wall at the lap. Your clapboard is tight to the sheathing at the top of the board (the thin part) and therefore is a true 4-foot 8-inch radius. At the bottom, however, the lap means you are bending around a 4-foot 8 1/4-inch radius. What happens is that the ends of the clapboard rise above the horizontal course line if you keep the board tight to the sheathing, or you get a gap between the sheathing and the top edge of the clapboard.

I have found a way to deal with this without soaking or steaming the clapboards: The idea is to cut the siding to a shape that will bend around the wall and still stay horizontal (use 6-inch-wide clapboards for this). To get the right shape, nail a starter strip around the bottom of the wall, then scribe a line around the wall at the height of the top of your 4-inch clapboard. Make a mark in the exact middle of a length of 6-inch clapboard and tack it to the wall with one nail so that its top edge touches the scribed layout line at the center mark. Then, with a person on each end of the clapboard, bend it around the wall so that its top edge is pressed flat to the wall. You will see that the ends fall below your horizontal line. Tack the clapboard to the wall so that each end is an equal distance below the line. Now scribe horizontal lines for the bottom and top of the clapboard. Remove the clapboard and cut to these lines; the clapboard will have a flattened U shape. Cut as many pieces as you need to that particular curve; the length of the pieces doesn't matter.

In cases where the length of the curved clapboards isn't great (for example, if you're siding between windows), it's possible to start with a curved board, and make a number of successively less curved boards, until, maybe 10 boards later, you are working with straight boards that are arching as you put them on. The change from curved to straight is gradual enough that no one can tell that your boards arch up off the horizontal line. At 10 boards from the soffit, you need to start cutting progressively more curved boards until the last one is the right shape and lies parallel to the soffit again. I did this recently on a three-story tower with 4-inch clapboards and I had to cut only 20 boards; the rest came straight out of the bundles. No one can see the arch unless I point it out.   — *Cyrus Miller*

Should I Set Siding Nails?

Q. When beveled wood siding is going to be painted, should the nails be set below the surface of the siding and filled, or driven flush with the surface of the siding?

A. Siding nails should be driven flush with the surface of the siding. Nails set and filled are very unforgiving of any movement or shrinkage in the siding or wall framing. This movement may cause "nail pops" — the exterior version of the infamous drywall nail pop. Hardboard siding should be fastened so the nail head is drawn "snug" against the siding.

On the other hand, I prefer to use finish nails to fasten exterior trim (the brick molding around a door unit, for example). The finish nails should be set and filled. DAP's linseed oil-based Painter's Putty (Dap Inc., 800/543-3840; www.dap.com) is my favorite, and it works well under an oil-based primer. I add "whiting" (a thickening powder available through paint suppliers) to this somewhat gooey putty to make it more workable. If a latex primer is used, it's important that the putty be allowed to dry for a few days before the primer is applied.

An exterior spackle, like UGL's 222 Spackling Paste (United Gilsonite Laboratories, 800/272-3235; www.ugl.com) can also be used to fill nail holes. Depending on the weather conditions, either oil or latex primer can generally be applied the same day over this product. Exterior spackle shrinks as it dries, but an experienced painter will allow for this shrinkage by "overloading" the hole that's being filled and sanding any proud material flush after it dries.

In situations where a latex primer will be applied immediately after the holes are filled, I would use a quick-hardening two-part automotive body filler.

I recommend hot-dipped galvanized finish nails for all trim work that will be filled. Wood siding should be fastened with stainless-steel ring-shanked siding nails. Though more expensive, stainless-steel nails are cheap insurance against bleeding and corrosion problems.

— Stephen Jordan

Using Pressure-Treated Wood for Trim

Q. We are planning to use pressure-treated wood as exterior trim on a house with fiber-cement siding. The siding will be finished with acrylic paint. Is pressure-treated wood appropriate for use as exterior trim? What precautions, if any, are necessary when painting pressure-treated wood? Can we use the same acrylic paint we will be using on the siding, or should we be using an oil-based paint?

A. Most pressure-treated wood sold in lumberyards today is treated with ACQ (alkaline copper Quarternary). Although this type of pressure-treated wood is paintable, be aware that painting is possible only when the wood has been cleaned (using soapy water and a stiff bristle brush, followed by a clear water rinse) and allowed to dry thoroughly. Getting the wood dry can sometimes be a problem, because treated wood is often sold very wet from the treating process. Depending on the climate and drying conditions, it may be necessary to dry the wood for several weeks before painting.

An exterior all-acrylic latex house paint would be the best choice for painting pressure-treated wood. Exterior acrylic latex house paints can normally be used on many different substrates — aluminum, galvanized steel, masonry, concrete, brick — as well as on pressure-treated wood and fiber-cement siding. However, always check the label on the paint can to be sure it is recommended for use on wood products.

If possible, find a manufacturer who also has an acrylic latex primer. The combination of latex primer and topcoat has been shown to give the best overall paint performance on treated wood. I would not use oil-based paint, which does not perform well on pressure-treated wood.

Pressure-treated wood may not be the best choice for exterior trim because most pressure-treated wood is southern yellow pine, a species that is not particularly good at holding paint. Southern yellow pine, whether or not it is pressure-treated, does not hold paint as well as western red cedar. Also, since most pressure-treated wood has knots and other defects, any lumber used for exterior trim would need to be carefully selected to find boards that are as clear as possible. Although some lumberyards do sell premium grades of pressure-treated wood for exterior trim, may be difficult to find.

Finally, since pressure-treated wood has a tendency to warp and crack rather easily, the trim would need to be carefully and securely nailed or screwed. — *Bill Feist*

Best Glue for Exterior Trim

Q. For gluing exterior wood trim components that will be exposed to the weather but not submerged under water, what's best: resorcinol, epoxy, or a polyurethane glue?

A. Any one of these will work, as long as the joint has good wood-to-wood contact. Personally, I think that epoxies and resorcinols have the best outdoor endurance qualities. For those joints that may be a bit loose, epoxy is the best, as it has the best structural and gap-filling qualities. For more information, go to www.homesteadfinishing.com. —*Jeff Jewitt*

Fiber-Cement With Aluminum Trim

Q. Is it okay to butt fiber-cement siding against aluminum-clad windows and aluminum soffit, or will the fiber-cement cause a reaction that damages the aluminum?

A. Portland cement is highly alkaline. Any cementitious material, including cement-based siding, can attack aluminum if the aluminum is not protected. Tamlyn & Sons, a manufacturer of vinyl and aluminum trim products, does not recommend that its aluminum trim products be used with fiber-cement siding. If you do choose to use aluminum trim, it should be anodized. Virtually all major window manufacturers anodize their aluminum trim. Aluminum windows used with stucco siding have a long history of successful performance.

To be safe, I would be sure that any cement-based siding is fully primed on all sides and the ends before installation, in order to minimize the alkaline bleed. Carefully prime the ends of the siding at field cuts. This will help minimize any possible reaction with aluminum. — *Paul Fisette*

No-Nails Approach to Aluminum Fascia

Q. Last winter, our siding sub installed aluminum fascia in very cold temperatures. When the weather warmed up, the fascia expanded and buckled severely. What's the proper way to install aluminum fascia to prevent this from happening?

A. To prevent the buckling you describe (photograph, below), fascia should be installed using a "nail-less" technique. By fastening F-channel to the bottom edge of the subfascia and utility trim under the drip-edge, the fascia material is held firmly in place but is still able to expand and contract with changes in temperature (illustration, opposite page).

Using a snap-lock punch, form locking tabs in the top edge of pre-formed fascia, then press the fascia into the utility trim. The tabs prevent the fascia from pulling out of the utility trim, but they don't restrict movement. Many contractors form their fascia on site using painted aluminum coil stock. Since coil stock is thinner than factory fascia (.019 versus .024 inch thick), you should form a continuous locking hem on the top edge and insert this into the utility trim.

To prevent the rake fascia from creeping downhill, the bottom edge of the fascia should be slot-punched, then held in place with three or four aluminum trim nails driven into the subfascia. These slots permit the fascia to expand and contract lengthwise without buckling.

This aluminum fascia, installed in cold temperatures with no allowance for movement, buckled when summer temperatures caused it to expand.

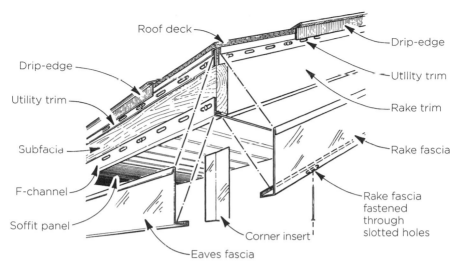

Roof deck

Drip-edge

Drip-edge

Utility trim

Utility trim

Rake trim

Subfacia

Rake fascia

F-channel

Soffit panel

Corner insert

Eaves fascia

Rake fascia fastened through slotted holes

Using stock profiles like F-channel and utility trim, aluminum fascia can be held in place without nails. When using aluminum coil stock, which is thinner than premanufactured fascia, site-bend a continuous locking hem on the top edge to help hold it snug in the utility trim. To keep rake fascia from sliding downhill, use aluminum trim nails driven through slotted holes in the bottom edge. The elongated holes allow the fascia to expand without buckling. Fascia should break at all inside and outside corners. A corner insert (center) prevents the framing from showing at the joint.

Finally, aluminum fascia should never turn a corner. Instead, fit a small insert at all inside and outside corners and break the fascia at these points. The insert provides a background that prevents the wood subfascia from showing at the corner joint. — *George Schambach*

Removing Vinyl Siding

Q. Can you explain the best way to unlock vinyl siding panels? We often run into vinyl when building an addition or cutting a new window opening, and we'd like to know how to remove and replace panels without destroying half a wall's worth.

A. Whenever I am called to a job to remove and replace vinyl siding, I follow these general steps:

I begin by looking over the job and figuring out which panels need to be removed. Rule of thumb: Always remove one course more than needed.

Starting at the highest course to be removed, I find the weakest or loosest end of the panel and separate it from the rest of the siding. While

Courtesy Malco

you can do this with the claw of a hammer, I recommend the "zip tool" (available from any vinyl siding supplier), which looks like a butter knife with a hook on one end (see photo). With this tool, you pry gently at the end of a panel, hook the lock, and pull down. The panel will come "unzipped" as you slide the hook along the lock.

To replace siding panels after the work is all done, follow these guidelines: Pay attention to the original butt lines and spacing. Lay up your panels and nail them in place. Make sure you don't overnail. The siding should be able to move under the head of the nail, so it can expand and contract with changes in temperature. If you pinch the panel tight, preventing it from moving, it will buckle.

Once you've installed the last course, close the seam by pulling down on the lock above with the zip tool in one hand, while gently pushing against the lock with your other hand. — *Mark Katuzney*

Does Wood Siding Need an Air Space Behind It?

Q. I'm considering installing wood siding over a rain screen, but I'm unsure whether the potential benefits justify the extra expense. What is the real benefit of an air space behind wood siding?

A. The main benefit of a rain screen is to increase moisture removal by creating a ventilated cladding. The ventilated cladding allows drying of both the siding (from the back surface) and the wall assembly (through the sheathing and building paper) into the air space behind the siding. Some sidings, including vinyl, aluminum, and brick veneer, are inherently self-ventilating. Other claddings, like wood siding, need help. In a rain-screen installation, the wood siding is installed over vertical battens. Alternatively, some installers vent clapboard siding by using wedges, clips, or oval-headed ("bumpy") nails to separate the siding laps.

What is the real benefit? Well, is it a benefit for the siding not to rot and the paint not to peel? Before the days of plywood, OSB, foam sheathings, cavity insulation, and interior poly vapor barriers, wet siding could dry toward the interior. Today's walls have low drying potential resulting from the use of impermeable or semipermeable sheathing, high levels of cavity insulation, and interior vapor barriers.

With foam sheathing, an air space is essential, both to reduce the siding's water uptake and to provide a receptor space for the moisture in the siding. If the cedar siding is installed directly on foil-faced foam sheathing, it will be prone to cupping, splitting, and premature paint peeling. In locations with high exposure to wind-driven rain, it may be

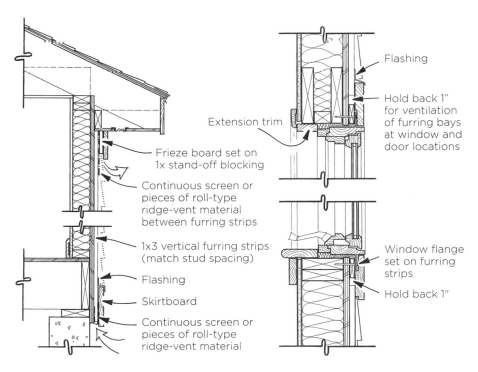

When applying wood siding over rigid-foam insulation, use furring strips to create an air space that buffers moisture cycling.

necessary to install tar paper or housewrap behind foam sheathing to help control any rainwater that penetrates the siding.

Finally, research has demonstrated that plastic housewraps and felt building papers can lose their water repellency when they are directly in contact with some types of wood siding because of tannins and other extractives from the wood. Similar problems may occur due to contact with soaps, detergents, bleaches, dirt, dust, and paint. None of this is a problem when the siding is installed over an air space. — *Joe Lstiburek*

Ghost Lines on Siding

Q. I'm working on a house where there are ghost lines of the wall studs on siding that is coated with a semitransparent stain. What might be causing this?

A. The ghost lines probably have little to do with the semitransparent stain. The lines are more likely from a thin film of dust (and possibly even mildew) that collects on the wall between the studs. The insulating value of the studs is considerably lower than that of the insulation-filled stud bays. The studs probably conduct enough heat to keep condensation from forming along them, while the siding between them becomes damp in times of high humidity. This microscopic film of condensation tends to hold dirt, which encourages mildew. The ghost lines will probably disappear if the siding is washed with a dilute bleach solution (a cup of household chlorine bleach to a gallon of water). To prevent the ghost lines, the siding would need to be isolated from the wall, either with foam sheathing or by installing the siding on strapping. — *Henry Spies*

Cutting and Patching Asbestos Siding

Q. Occasionally, we run across asbestos shingles. We sometimes have to cut the material for a door or window opening. What precautions need to be taken when cutting through this material? What is the best way to dispose of a few shingles? Is this material still available for repairs?

A. Cutting or breaking asbestos cement board or shingles releases asbestos fibers into the air. To remove shingles without breaking, apply pressure near the nail head to expose the head enough to cut it off with side-cutters.

If at all possible, asbestos shingles should be cut with a shingle cutter. Sawing siding panels with an abrasive saw will release large quantities of fibers. The shingles can also be scored with a carbide knife and snapped clean. In either case, a respirator with a HEPA filter should be worn.

There is controversy over whether incidental exposure to chrysotile (white) asbestos, the type used in most residential work, presents a significant danger. When intact and in good condition, asphalt or cement siding or roofing materials that contain asbestos are generally considered "nonfriable" and not generally hazardous if left alone. (*Friable* means that the material can be broken up by hand pressure.) Extended exposure to any of the mineral fibers, however, is very hazardous.

Be sure to familiarize yourself with your state's asbestos removal laws before proceeding. The requirements for removal and disposal vary from state to state. Depending on the type and amount of asbestos-containing material involved and the usage of the building, many states require that the material be removed by a certified or licensed asbestos worker. Some states require that the material be enclosed in a crate or heavy plastic bag and be buried in a hazardous waste landfill. Other states only require that it be bagged and marked, but buried in the ordinary landfill.

Salvage asbestos tile may be available from Slate Roof Central (814/786-9085, www.jenkinsslate.com)   — *Henry Spies*

Preventing Stucco Cracks

Q. We often get conflicting information from subs about certain stucco details. Specifically, what's the best way to avoid cracking? Must the scratch coat cure for 48 hours before applying the brown coat, or can the brown coat go on the same day? How long should the scratch coat cure before the color coat? Also, does one type of lath perform better than another?

A. On any stucco job, you have to expect both the scratch coat and the brown coats to have minor cracking from shrinking. Excessive cracking, however, usually means something was done improperly — for example, the mix may have been "too rich" (too much cement), the stucco layers may have been too thick, or the walls may not have been allowed to cure properly.

A "too rich" mix almost always results in cracking, especially when the coat is applied too thickly. Normally, the scratch coat should just cover the wire. The next layer (the brown coat) may be applied the next day. Ideally, both surfaces should be "misted" after they've been allowed to set awhile. This will slow their curing and maximize the strength of each coat. It's especially important to "damp cure" the walls on very hot and windy days. If either coat flash-cures, it will be weak.

You should allow a week to pass before finishing the stucco. This permits complete curing of both the scratch and brown coats. The finish coat (about 1/8 inch thick) is applied over the previous two coats (totaling about 3/8 inch), so any hairline cracks become covered. Nevertheless, stucco is a cementitious material, so it is very unforgiving. Even if there are no cracks at all in the scratch and brown coats when the finish is applied, you will undoubtedly find some cracking several months later because of expansion and contraction of the framing, movement caused by wind loads, and, of course, any seismic activity.

As for lath, here in central Ohio we've had good results with 1-inch, 18-gauge, key-mesh wire lath, otherwise known as self-furring metal lath or stucco netting. For overhangs, we use expanded metal lath (sometimes with paper backing) because it's easier to handle when working overhead.

— *Steve Thomas*

Installing New Stucco Over Old

Q. The owner of a house with existing exterior cement stucco is unhappy with the finish. How difficult is it to install a new coat of stucco over the existing stucco?

A. Applying new stucco over old stucco (often called re-stuccoing) is a fairly simple process. If the existing wall is in good condition, this job can be straightforward. But if the existing surface has imperfections, the problems that caused the flaws must be repaired or else the problems will recur. Before re-stuccoing, thoroughly examine the existing surface and ask the following questions:

- Is there any loose, spalling stucco? Rub and tap the wall, listening for hollow sounds. Any loose stucco will need to be removed by scraping or sandblasting. To patch the stucco, combine sand and cement with calcium aluminate, an accelerator, or use a good non-shrinking stucco patching material or a rapid-set mortar mix.
- If the house has been painted, is the paint in good condition? Loose or chipping paint can reduce the bonding power of a new coat of stucco, so it should be removed by sandblasting. If the painted surface is in very good condition, you can apply stucco directly over it, as long as you use a bonder.
- Is there efflorescence? Efflorescence is a white powder or film on a surface, composed of salt crystals left behind when salt-laden water evaporates. Efflorescence reduces the bonding power of the new coat of stucco. To neutralize efflorescence, spray on a mild acid such as vinegar. Let the vinegar sit for half an hour, and then flush the wall with water. This will bring the pH down to an acceptable level.
- Is the wall dirty? Dirt, like loose paint or efflorescence, will interfere with the bond of the new stucco. Any dirt should be washed off.
- Does the wall have any cracks or leaks? Leaks near doors and windows, as well as cracks, should be repaired before re-stuccoing.
- Is the flashing in good condition? Inspect the metal flashing and weep screed for rust or separation at the joints, and repair or replace it as necessary.
- Is the existing stucco surface rough and uneven? If so, scrape down the high points and fill in the low areas before proceeding.

Once all problems in the existing stucco have been corrected (including, if necessary, adding a leveling coat to even out the low areas), apply a new finish coat with the texture and integral color of your choice. To ensure a good bond, use a bonder between the existing stucco and the new material.

— *Ron Webber*

Stucco Over Concrete

Q. I'd like to finish the above-grade portion of a poured-concrete foundation with a color coat of stucco. The broken-off snap-ties are visible and are slightly rusty. Do I need to treat these with anything before I apply the stucco to prevent a rust stain on the finished surface?

A. Ideally, the snap-ties should be broken back so that the ends are 1 to 1¹/2 inches from the surface of the concrete. Then the holes around the snap-ties should be plugged with hydraulic cement to prevent moisture from entering through capillary action. This should prevent rust from bleeding through and staining the stucco coat. It sounds in your case like the concrete contractor used the kind of snap-ties that are designed to break off at the surface of the concrete. These are fine if the intention is to cover the foundation wall with a drain board or foam insulation system. But there are also special ties available that will break off below the surface. Next time you plan to stucco a wall, let your foundation contractor know in advance, and he can use those.

I've been in your situation a couple of times. Once, I drove a piece of ¹/4-inch tubing as far over the ties as possible and used it to rock the ties back and forth until they snapped off below the surface of the wall. Another time, I used a grinder to take them down. This also removed some of the concrete around the tie, but the hydraulic cement patched that. Fortunately, you usually have to stucco-coat only a narrow band of wall that's exposed above grade, so you won't have to treat every tie in the wall.

— *Jay Meunier*

Patching Stucco

Q. When making a patch in stucco, is it possible to use hydraulic cement or a fast-setting cement for the first couple of layers so that the patch can be completed in one day?

A. I would advise against either hydraulic cement or any "hurry-up" product for the base coats. The reason is simple: You can almost bet the surface you're attempting to patch (and match) was not created using these products but a less expensive, standard cement/lime/sand mixture.

The capillary suction of the undercoats — their tendency to absorb moisture from the finish coat — will directly affect the drying time and therefore the color of the top coat. The hard, fast-drying products you refer to are much less absorptive than a standard cement mix. Even if you use the exact same finish-coat recipe as the original, it will end up looking different over a fast-setting undercoat. Typically, the faster the stucco dries, the lighter the color will be.

If possible, seek out the original stucco contractor and try to get the recipe for his finish mix, to get you as close as you can to a perfect match. Even so, you'll have some difficult hurdles to overcome. Age and weathering tend to darken stucco colors. Your best bet, depending on the size of the patch, may be to repaint that entire side of the house.

In general, it's not a good idea to rush stucco work. Put on the scratch coat one day, let it cure overnight, then put on the brown coat. The brown coat should be flat and held consistently about 1/8 inch below the surrounding finish coat. Irregularites in the brown coat will cause the finish coat to dry unevenly, and will produce a mottled color as well as excessive cracking. For best results, leave the brown coat for a week before putting on the finish coat. This will minimize cracks in the top coat.

— *Steve Thomas*

Flashing a Stucco Chimney

Q. What is the best way to flash around a wood-frame-and-stucco chimney for asphalt shingle roofs?

A. The best way is to build and sheathe the chimney, then flash to the sheathing with the same step flashing that's recommended for a brick chimney. Next, install waterproof building paper or peel-and-stick membrane over the sheathing and flashing as a counterflashing (see illustration at right). Attach the metal lath or welded wire lath with furring nails, which keep the metal spaced away from the waterproof layer.

Apply the three coats of stucco, and finish the bottom with a stop screed, or weep screed, about 1/2 inch above the shingles. The step flashing should show enough to allow the application of a second layer of asphalt shingles in the future without touching the screed.   — *Henry Spies*

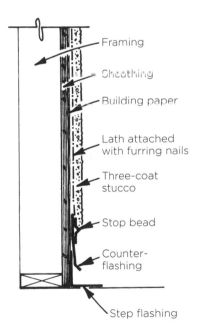

Framing

Sheathing

Building paper

Lath attached with furring nails

Three-coat stucco

Stop bead

Counter-flashing

Step flashing

Flash a stucco chimney with step flashing, just as you would a masonry chimney. Install a stop screed to finish off the stucco edge, and place it high enough to allow for a second layer of roofing in the future.

Repairing a Stucco Corner

Q. What's the best way to build up a damaged bullnose corner on exterior stucco?

A. Repairing a damaged rounded corner is handled like any other repair job. Assuming that the corner has been hit and a chunk of stucco has been knocked out, the loose and damaged stucco should be chipped back. Carefully tuck the top of the weather barrier patch under the existing weather barrier, creating a lap, so that the weather barrier keeps out the water.

Then install the wire or metal lath, again lapping the old and the new by about 2 inches. If there is nothing to nail to at the repair location, then you can tie the lath together using tie wire. Now you're ready to install the stucco base coat, taking care to maintain the desired shape. If it is a heavy fill, you may need to do it in two coats. The finish coat should be installed using the same technique that the original plasterer used, in order to match the texture.   — *Ron Webber*

Wood Trim With Stucco

Q. We will soon be building a stuccoed Tudor-style home. Standard practice in our region is to fasten the wood trim directly to the sheathing, then apply a three-coat stucco to the areas formed by the applied trim. How can the trim-stucco joint be detailed to prevent water infiltration?

A. The key to leak-free details is to install "build-out boards" behind the trim. In our area, 1x6 cedar is commonly used for Tudor trim. We center the vertical and angled trim boards over a 1x4 build-out board (see illustration). The smaller-width build-out board allows the two stucco base coats to be troweled in behind the 1x6. The finish stucco coat is run tight to the edge of the overlapping 1x6 trim board, which discourages water from working its way behind the base coats.

Unlike vertical trim, horizontal trim is held flush with the upper edges of the build-out boards, and a galvanized cap flashing is installed to prevent water infiltration.

It's important to follow the proper sequence when installing these Tudor trim assemblies. Horizontal assemblies and cap flashings are installed first, housewrap or Class D building paper is stapled to the sheathing, and then vertical and angled trim assemblies are installed.

If cedar trim is used, insist that it be prestained on all sides before

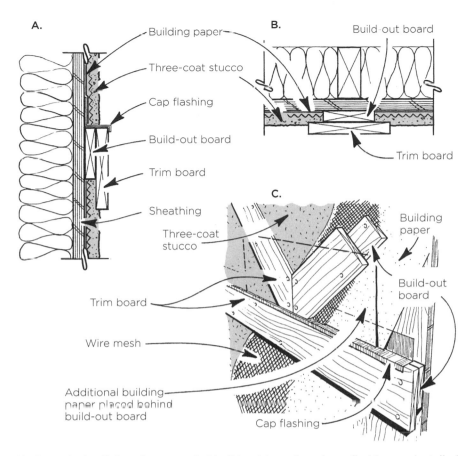

A.
- Building paper
- Three-coat stucco
- Cap flashing
- Build-out board
- Trim board
- Sheathing
- Three-coat stucco
- Trim board
- Wire mesh
- Additional building paper placed behind build-out board

B.
- Build-out board
- Trim board

C.
- Building paper
- Build-out board
- Cap flashing

Horizontal trim (1x6 cedar over a 1x4 build-out board) and cap flashing are installed first, directly over the sheathing, followed by cap flashing (A). Next, housewrap or Class D building paper is installed, lapped carefully over the cap flashing. Vertical (B) and diagonal (C) trim assemblies follow. An extra layer of building paper behind the bottoms of diagonal trim protects against the additional water that funnels to those spots.

installation, and that all site-cut ends have stain applied to them. Cedar contains extractives that can "bleed out" if not properly sealed.

— *Steve Thomas*

Shrinkage Joints in EIFS Cladding

Q. We're building a three-story house with plywood sheathing that will be covered with EIFS. Because the house has 9-foot ceilings, the horizontal sheathing joints don't fall near the band-joist area. The EIFS rep says we have to plan for joist shrinkage by establishing a control joint, and he wants us to cut through the plywood sheathing at the floor line. I think the plywood sheathing is structurally important, and I'm reluctant to cut the sheathing where it ties the floors together. What should we do?

A. When EIFS is installed on a wood-frame building, EIFS manufacturers recommend the installation of a horizontal joint in the sheathing at floor lines. The purpose of the joint is to accommodate the cross-grain shrinkage in the floor joists, as well as any settling. (This type of shrinkage can cause problems with many types of cladding, including EIFS, brick veneer, and vinyl siding.) In addition to providing a gap in the sheathing, a flexible joint is provided in the EIFS cladding to accommodate the movement that occurs as the sheathing joint closes.

This standard detail is problematic when the sheathing spans the floor line to structurally tie the upper floor to the lower floor, as is often the case in seismic zones and high-wind coastal areas. In such cases, the EIFS joint should be installed at the nearest horizontal joint in the sheathing, even if that is not at the floor line. As the joists shrink, the stress will probably be focused at that sheathing joint.

This is the best solution, but it is not ideal. The stresses are somewhat unpredictable — if the sheathing is securely restrained above and below the floor line and the shrinkage is severe, the sheathing may bulge as it is compressed, causing the EIFS also to bulge out or crush at that point.

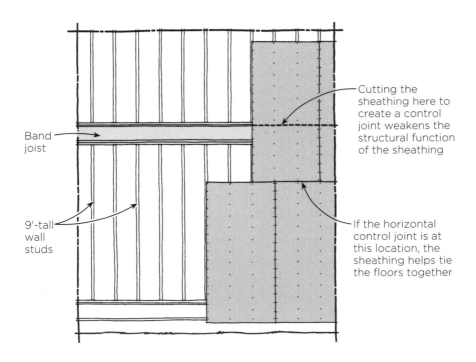

Band joist

Cutting the sheathing here to create a control joint weakens the structural function of the sheathing

9'-tall wall studs

If the horizontal control joint is at this location, the sheathing helps tie the floors together

More information can be found in the EIMA *Guide to EIFS Construction,* which is posted at the Web site of the EIFS Industry Members Association (www.eima.com). — *John Edgar*

Should EIFS Extend to Grade?

Q. Is it acceptable to terminate EIFS at grade? I'm worried that this might provide easy access for termites, or might allow moisture to wick into the wall.

A. EIFS should be terminated above grade for the two concerns you mention: termites and the possibility of moisture entering the wall by capillary action. Snow and ice can also cause water to back up into the wall in winter. Most building codes require the top of the foundation to be a minimum of 6 or 8 inches above grade for the same reasons. If you want the color and texture of the EIFS finish to extend to grade, the exposed portion of the foundation can be skim-coated with EIFS base material and finish.

Ideally, the lower edge of the EIFS extends approximately one inch below the top of the foundation and is sealed to the foundation to prevent air infiltration. If the resulting thermal loss through the exposed foundation is your primary concern, it is better to address it from the inside, even though interior foundation insulation is slightly less effective than exterior insulation. The possible damage by moisture and termites outweighs, in my opinion, any advantage of continuous exterior thermal insulation. Also, completely covering the exterior of the exposed foundation is prohibited by some state codes and violates many exterminator warranties. — *Richard Piper*

Mounting Downspouts on EIFS

Q. What is the best way to attach gutter downspouts to a house with EIFS?

A. To fasten something light to an EIFS wall, you will need screws or lag bolts long enough to reach through the rigid foam to the sheathing or framing. You'll also need a PVC sleeve (plastic tubing or conduit) with an interior diameter a little larger than the fastener. The sleeve allows the installer to tighten a screw against a solid object without crushing the EIFS.

Fastening an object through EIFS will be easier if the sheathing under the foam is plywood rather than gypsum. If the sheathing is plywood, use the following procedure:

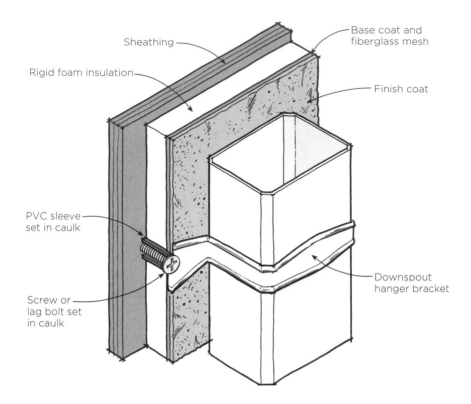

Sheathing

Base coat and fiberglass mesh

Rigid foam insulation

Finish coat

PVC sleeve set in caulk

Downspout hanger bracket

Screw or lag bolt set in caulk

1. Mark the location of the fastener on the finish.
2. Drill a hole through the EIFS up to — not through — the sheathing.
3. Line the hole with a neutral-cure caulk. (If the caulk smells like vinegar, it should not be used. A vinegar smell indicates an acid-cure caulk, which may corrode the fastener.)
4. Cut a length of your PVC sleeve about 1/8 inch longer than the thickness of the EIFS, and insert it into the hole. The sleeve should be 1/8 inch proud of the EIFS. Tool any excess caulk.
5. Fill the sleeve with additional caulk and immediately fasten the downspout through the fresh caulk in the sleeve. Remove any excess caulk.

If the sheathing is gypsum, the only locations where fastening is possible are at studs or blocking. Follow the same procedure described above. If it is necessary to fasten between studs, a toggle bolt may work. Of course, before drilling through any wall, be sure that there are no electric wires or gas lines in the area. — *John Edgar*

Matching Old Mortar

Q. In an upcoming remodeling project, we will be uncovering a section of hidden stonework that will need to be repointed to match the existing exposed stonework. How do we go about matching the color of the mortar?

A. When matching existing mortar, I first try to decide whether a match can be achieved using "standard" mixing ingredients. The pointing mix I use consists of one part Portland cement and three parts sand. Portland cement is available in gray or white, and sand in my area is available in white, yellow, or brown. When gray cement is used in the pointing mix, the color of the sand has little effect on the final color of the mortar. When using white cement, you can control the color with the sand used in the mix. White sand will produce a white mortar, yellow sand a beige mortar, and brown sand a light brown mortar with a reddish tint.

In my area of Pennsylvania, I've had the best success matching the mortar found in older stonework by using white Portland cement and a mixture of brown and yellow sand. Before I begin pointing, I premix all the sand needed for the job. The premixed sand can then be mixed 3 to 1 with the cement.

If you can't match the mortar using readily available materials, you'll have to use solid-color dyes. These are packaged dry and must be mixed in thoroughly when the sand and cement are dry-mixed. You'll need to make numerous test batches and allow them to dry for about two days before comparing them with the existing mortar. However, after the dry materials for each test batch are thoroughly mixed (and before any water is added), you'll have a close indication of the final color.

Dry dyes tend to be very concentrated: A one- or two-pound bag will tint a full 94-pound bag of cement. If only one or two masons are pointing, a full bag mix is too much mortar to mix at one time. When making smaller batches, you'll need to carefully control the amounts used in the mix to maintain a consistent color. It's important to finish your test samples the same way you'll be finishing the final work (e.g., a brushed joint will have a different color than a smooth troweled joint). — *Tony Jucewicz*

Reusing Brick

Q. When rebuilding a chimney above the roofline, is it okay to reuse the existing bricks? I've heard that mortar doesn't stick well to reused bricks.

A. You are correct — salvaged bricks normally develop low bond strength when used in new construction. When a new brick is placed in mortar, its pores absorb the water in the mortar, drawing in the cementitious material. Once the mortar has cured, this cementitious material will bond the brick to the mortar, creating a solid wall. In salvaged brick, the pores of the brick are, in essence, clogged by the hardened mortar materials, preventing it from developing the same bond strength as new brick. So, I would be extremely cautious in using salvaged brick. For more information, visit www.brickinfo.org. — *Richard Allen, Jr.*

Weep Holes in Brick Veneer

Q. Are weep holes in a typical wood-frame brick-veneer home required anywhere other than at the bottom? What about over and under windows? Also, is it required that a brick windowsill be pitched? Are weep holes required in faux-stone installations?

A. Yes to all questions: it's important to bring any water that might be running down the surface of the wood-framed wall back out on the surface of the brick anywhere it might enter the framing. The 2003 IRC requires minimum 3/16-inch-diameter weep holes every 33 inches, just above the flashing (R703.7.6). Flashing, in turn, is required under the first course of masonry at ground level, above windows and doors, below window sills, and at any lintels and shelf angles (R703.7.5, R703.8). Many of these details are included in Figure R703.7. The Brick Industry Association (www.bia.org) is an excellent source of information on proper brick-veneer construction; the illustration (opposite) is based primarily on BIA recommendations, which frequently go beyond code minimums. Oddly, the IRC doesn't require building paper over the plywood or OSB sheathing as long as there's a 1-inch air space. However, both the BIA and the APA recommend paper, and it shouldn't be left out. — *Don Jackson*

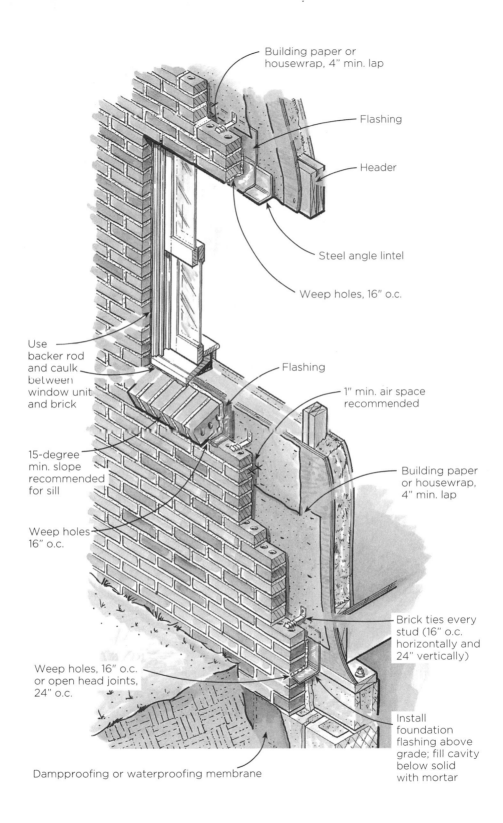

Building paper or housewrap, 4" min. lap

Flashing

Header

Steel angle lintel

Weep holes, 16" o.c.

Use backer rod and caulk between window unit and brick

Flashing

1" min. air space recommended

15-degree min. slope recommended for sill

Weep holes 16" o.c.

Building paper or housewrap, 4" min. lap

Weep holes, 16" o.c. or open head joints, 24" o.c.

Brick ties every stud (16" o.c. horizontally and 24" vertically)

Install foundation flashing above grade; fill cavity below solid with mortar

Dampproofing or waterproofing membrane

Overhanging Brick Veneer

Q. It's not unusual for a foundation to be slightly out of square. If the house has brick veneer siding, it's sometimes necessary for the first course of bricks to overhang the concrete foundation. What is the maximum safe overhang in such a situation?

A. To provide structural stability, at least two-thirds of a brick masonry wythe should bear on the foundation (see illustration). Three-inch brick veneer should not overhang the foundation by more than 1 inch, and $3^5/8$-inch brick veneer by not more than $1^1/4$ inches.

One reason this condition occurs fairly often is that the American Concrete Institute's *Standard Tolerances for Concrete Construction and Materials* (ACI 117) provides that footings may be misplaced as much as plus or minus 2 inches. One solution to a misplaced footing is to relocate the wall. The *Specifications for Masonry Structures* written by the Masonry Standards Joint Committee (MSJC99) provides that the location of walls may differ from the intended location by as much as plus or minus $3/4$ inch. So as long as the wall location can be adjusted in the right direction, it is possible for $3^5/8$-inch brick veneer to stay within the allowable tolerances for both masonry and concrete. — *Clayford Grimm*

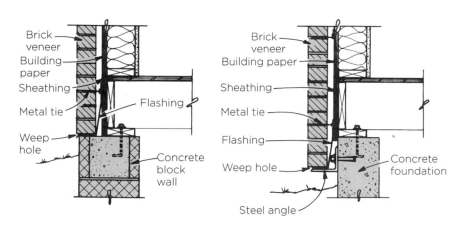

In new construction, build foundation walls wide enough to provide a 4-in. shelf for the brick (left). A stainless-steel angle bolted onto an existing foundation (right) can support a retrofit brick veneer wall up to 14 ft. high. The steel should be installed above-grade, and should be wide enough to support two-thirds the width of the brick.

Pressure-Washing Before Painting

Q. Is it a good idea to pressure-wash wood siding before painting? It seems like this would drive a lot of water into and behind the siding — water that might still be present when the paint is applied. And what about pressure-washing decks?

A. Pressure-washing is a good way to remove dirt and loose paint, but you must use the pressure-washer carefully. Add a mild degreaser such as TSP to the water, and wait a day or two for the walls to dry out before applying paint. Many modern latex paints actually allow moisture to breathe out through their dried film. If using oil paints or stains, let the surface dry for an extra day or two before application. The same holds true for decks.

Remember that it is easy to do a lot of damage in a hurry with a pressure-washer, especially to wood siding and decks. Improperly handled, a pressure-washer can gouge soft wood, drive water into the interior of buildings (staining walls and ceilings), break glass windows and light fixtures, and injure the operator or someone nearby.

When pressure-washing wood siding that has peeling paint, it is important to let the entire surface dry and then scrape by hand. This is because the moisture will also start to lift the paint around the paint that was peeling. You don't want to paint over this compromised surface, because it will be the first thing to fail.

In California, all deck stains and sealers have been reformulated to meet tough new pollution standards, and this has affected surface preparation. Many manufacturers now recommend that decks be carefully pressure-washed in order to remove all of the previous finish and to open up the pores of the wood to accept the new sealer. Then look forward to repeating the process every couple of years. — *James Benney*

Staining New Boards to Look Old

Q. We are building an addition on a 25-year-old house. The house has cathedral ceilings with exposed unfinished pine board roof sheathing, and the walls are paneled with unfinished pine boards. The pine boards have darkened with age, and the homeowners want the new pine boards in the addition to match the existing boards. What's the best way to achieve this?

A. The color change in the old wood is caused by ultraviolet and visible light. Since the color change is just on the surface, it could be removed by sanding. Of course, sanding all of the existing roof sheathing and paneling to reveal the blond wood underneath would be difficult and tedious, so the best way to achieve a match is to stain the new wood.

That said, it is difficult to stain new pine to exactly match the color of wood that has been slowly changing for 25 years. You should warn the homeowners that an exact match will not be possible. One trick is to thin out some of the stain used on the new wood to also color the old wood — but only slightly.

Over time, the color of the new stained wood will most likely change, but the stain pigments, by blocking some of the light reaching the wood, should slow down that process. Hopefully, with time, the old and the new wood should look nearly alike.   — Bill Feist

How to Prime New Pine

Q. We're new home builders and use a lot of newly milled kiln-dried eastern white pine exterior trim. We paint it in batches, spraying both sides with a good-quality oil-based primer, and then follow with two topcoats of 100% acrylic. To cut down drying time, we'd like to switch to an acrylic primer, but most painting contractors we talk to swear by an oil-based primer for pine because it penetrates better. Is there an acrylic primer that would give good performance?

A. Why use oil primer first? New pine has a hard surface that allows little penetration of an oil primer.

I recommend using a latex primer first, followed by a second coat of oil primer if your customer insists. But you could also use two coats of latex primer, if you prefer. There are many good acrylic primers that could work for you. California Paints and Sherwin-Williams both make a 100% acrylic. I happen to like Muralo's Cedar Solution, which has good tannin-blocking properties, and Muralo's 2205 Universal 100% acrylic primer; both have excellent adhesive properties (800/631-3440, www.muralo.com).

There are other good products on the market; it is really your choice as the finisher. Many universal latex primers are designed for smooth surfaces, which make them a good match for new pine. It's always a good idea to cut the pine resin by sanding with 80- or 100-grit paper. You should also ease the edges, because paint doesn't stick well to sharp corners.

Note that I recommend two coats of primer, not one. If you are putting on three coats total, I would prime twice and top-coat once. The primer coat is where adhesion to the substrate has to be perfect. The topcoat doesn't do much good if the primer comes off. You only get one chance to get the primer right. If you're spraying primer, you should always back-brush, so as to fill the grain of the wood. If you simply spray the primer, it will sit on the surface and adhesion will not be as good.

— *Duffy Hoffman*

Sealing Exterior Wood Before Painting

Q. How much benefit is there to applying a water sealer and wood preservative to clapboard siding before priming and painting?

A. There are many advantages to using a paintable water-repellent preservative (WRP) on unpainted clapboard siding before priming and painting. The treatment reduces raised grain, checking, warping, and splitting, and also improves paint adhesion. A WRP inhibits mildew growth on both painted and unpainted wood, and will retard decay in above-ground applications. A paintable preservative will help improve paint performance on the more difficult-to-paint woods, like flat-grain southern yellow pine and other flat-grain wood species.

Use WRPs only on dry, bare wood. The treatment must be done when the temperature is above 50°F. You can apply the treatment by brush or by dipping. When brushing, allow two days of warm, favorable drying weather before painting. When dipping, a full week of favorable drying weather may be necessary before painting. If you don't allow enough time for most of the solvent to dry from the wood and for the wax in the preservative to be absorbed, the paint may not cure or bond properly.

If you want to apply a WRP to previously painted wood, remove all loose paint, then brush the preservative into joints and unpainted areas only. Remove excess WRP from the painted surfaces with a rag. Again, allow two days of favorable warm drying weather before repainting.

Commercially available WRPs include Clear Wood Preservative (in the Cuprinol Group) from Sherwin-Williams and DAP's Woodlife Classic II Wood Preservative (www.sherwin-williams.com; www.dap.com). When shopping for a WRP, the key word to look for is "paintable." There are many WRPs that are meant to be used as wood deck treatments and natural finishes, but these are generally not paintable.

— *Bill Feist*

Paint vs. Stain on Clapboards

Q. I plan to install radially cut spruce clapboards on a new house and want to stain them with a white semitransparent or opaque stain. What type of finish appearance should I expect, and how often will the finish need to be renewed? Although stained siding looks better to me than painted siding, I know that some people say that stain is "just thinned paint."

A. Semitransparent stains are most effective on rough-sawn and weathered wood because more finish can be applied. The best exterior house stains are usually described as semitransparent, penetrating, and oil- or alkyd-based (solvent-borne). The better alkyd-based penetrating stains contain a fungicide (preservative or mildewcide), an ultraviolet light stabilizer, or a water repellent. Check the label for these important ingredients.

The alkyd-based solvent-borne stains actually penetrate the wood surface to a degree, and do not form a surface film like paint does. Thus, they don't totally hide the wood grain and will leave a soft, flat appearance. They will not trap moisture that may encourage decay. Since they penetrate and don't form a film like paint does, the stains cannot blister or peel even if moisture penetrates the wood. Alkyd-based stains normally only require a light cleaning with a stiff bristle brush and water before refinishing.

Latex-based (waterborne) stains are also available, but they do not penetrate the wood surface as do their oil- and alkyd-based counterparts. These are essentially "thinned paints." On spruce clapboards, a latex stain probably won't perform as well as an oil- or alkyd-based stain, and could also be more difficult to refinish.

How long the stain will last depends on weather exposure and the roughness of the wood. When used on new smooth-planed siding that is fully exposed to the weather, semitransparent penetrating stains generally last only about two to three years. When refinished after weathering, a smooth-planed siding should accept two coats of stain, and the finish will usually last much longer than the first application. Since a rough surface will usually accept two coats of stain, even on the first application, it is preferable to a smooth surface. Stain on rough-sawn siding may last six to eight years, depending on the amount of exposure. However, such durability often requires applying the stain at a rate of 100 to 150 square feet per gallon, a much greater amount than usually required for paint, which is typically applied at a rate of 400 to 450 square feet per gallon.

— Bill Feist

Preserving Seaside Shakes

Q. What is the best finish for the cedar shakes on a coastal home that is exposed to strong sun and lots of fine salt spray? Should the finish be sprayed, dipped, or brushed for best results?

A. Much depends upon the desired look. If you want a weathered gray look that is common in coastal areas, a clear water-repellent preservative can be used as a natural finish. It reduces warping, prevents water staining at the edges and ends, and helps control mildew growth. However, it does not, and cannot, maintain the new look of the cedar.

If you want the gray look immediately, a preservative stain is the answer. Semitransparent stains contain inorganic pigments, water repellents, and preservatives (mildewcides and fungicides). A semitransparent stain is more durable than a clear water-repellent preservative because the pigment absorbs some of the ultraviolet light, which deteriorates the wood fibers. Solvent-based (oil-based) stains are preferable to latex stains because they penetrate the wood surface and aren't as likely to blister or peel. Latex-based stains and solid-color stains (whether latex- or oil-based) form a thicker surface film. While this film will guard against ultraviolet light, it may also peel or flake.

Dipping the shakes will provide the best coverage and will seal their cut ends, sides, and backs, reducing warping. Dip the shingles to slightly more than double the exposure, so the wood exposed in the cracks between the shakes will also be treated. — *Henry Spies*

Painting Fiber-Cement Siding

Q. Fiber-cement siding comes primed. Can I go straight to a top coat? What's the best paint to use?

A. There are a couple of potential problems with the primer that comes on fiber-cement board. First, it's probably been thinned to make it easier to spray, which dilutes both the primer and the mildewcide in it. Also, you don't know how long ago the material was primed. Primer holds its tooth for only 30 to 60 days; after that, the surface should be reprepped.

Unless you have reliable information about when and how the material was primed, I would err on the side of caution. A good substrate ensures a good top coat. If the substrate fails, so will the top coat. Even if you could get the siding manufacturer to cover the cost of the paint, it wouldn't cover your labor cost to scrape, sand, and recoat. Because fiber cement is a hazardous material to sand, to me it makes more sense to prep the substrate correctly in the first place.

Here's what I would do. First, wash the siding with Pittsburgh Paint's Mildew Check (www.ppg.com). This is better than using a bleach solution, which dissipates within about 48 hours, allowing mildew spores to once again begin growing. Mildew Check leaves a longer-lasting film of mildewcide on the surface. Next, I would lightly etch the surface with 150- to 320-grit sandpaper, then prime with a good acrylic primer, followed by a 100% acrylic top coat. You can also buy unprimed fiber-cement siding and follow the same steps.   — *Duffy Hoffman*

Painting Aluminum Siding

Q. Is it possible to paint aluminum siding?

A. Yes. If you carefully apply one or two coats of premium 100% acrylic latex paint on aluminum siding, you can expect it to last for about 10 years.

Before painting, remove any chalk, dirt, or mildew by thoroughly washing the siding, preferably with a power washer. Aluminum siding, especially if it has been directly exposed to sun and rain, may be heavily chalked. If there are any signs of mildew, a mildewcide (available from your paint dealer) should be added to the paint. In order to prevent mildew from recurring, it's best to paint the siding as soon as possible after cleaning.   — *Ed Fillbach*

Clapboard Paint Problems

Q. The red cedar clapboard siding on a house in the Boston area was not back-primed and has typical moisture-related paint failure. It has been painted five times in 25 years. The house has no interior vapor barrier. I am considering removing the siding and installing new back-primed clapboards over a vented rainscreen. Will the rainscreen installation solve the problem of premature paint failure, or must I address the lack of an interior vapor barrier?

A. Installing new back-primed clapboards over a rainscreen will almost certainly solve the problem of premature paint failure, regardless of whether the house has an interior vapor barrier. A rainscreen combined with plenty of mechanical ventilation and air-sealing by a reputable weatherization company will go a lot further to alleviate indoor moisture problems that might be affecting the paint job. In addition, the home-owner should know that every house needs adequate mechanical ventilation, both to maintain indoor air quality and to reduce the chance that interior humidity could migrate into the walls.   — *Mark Snyder*

Keeping Cedar Siding Looking New

Q. What's the best method to keep cedar siding and trim looking naturally orange and new?

A. Unless you shrink-wrap each piece of cedar, it will darken and eventually weather to a silvery gray, or worse, to a blotchy dark gray. Clear finishes with UV inhibitors will slow down the weathering process, but these still need to be reapplied every two years or so. Eventually the color will change anyway, even indoors.

The best thing you can do is to approximate the color of new wood by putting on a cedar-tone stain. First, install the siding with the rough side out. A smooth surface holds less finish and weathers much more quickly than a rough-sawn surface. Finish the rough surface with two coats of a lightly pigmented, semitransparent, oil-based stain. Be sure to choose one that contains a water repellent and a preservative or mildewcide for best performance. Apply the first coat and let it soak into the wood 20 to 60 minutes and then apply the second coat. If you allow the first coat to dry, the second coat cannot penetrate into the wood. About an hour after applying the second coat, use a cloth, sponge, or dry brush to remove any excess stain. Otherwise, the stain that does not penetrate into the wood will form an unsightly film and glossy spots. Two coats of oil-based stain on rough wood will last from four to eight years, depending on the weather conditions it is exposed to.

If you have weathered and discolored wood siding, you can regain the new look of cedar by cleaning off the dirt and mildew with a solution of one-third cup liquid household detergent (be sure it is ammonia-free), one quart liquid household bleach (containing 5% sodium hypochlorite), and three quarts warm water. Follow this up with a water rinse and then use an oxalic acid bleach solution made with about a half pound of oxalic acid per gallon of water. Be sure to rinse with water again. This oxalic acid bleach solution will draw out the tannins in the wood and revive the orangeish tone of the cedar. At this point, you can let the wood weather naturally, or apply the cedar-tone semitransparent stain.   — *Bill Feist*

Painting PT Decking

Q. Some of my customers want their pressure-treated decks painted, but I've had trouble getting paint to adhere well to pressure-treated wood. What's the solution?

A. Treated lumber is not the primary source of your deck-painting difficulties. Rather, a paint's performance suffers on a horizontal surface that is exposed to the weather. To make matters worse, deck boards are usually flat-grained, high-density wood that doesn't hold paint as well as edge-grained, low-density wood.

For exterior decks, you're better off using a water-repellent preservative or a penetrating-type semitransparent pigmented stain. Film-forming finishes such as paints and solid-color stains aren't recommended for horizontal surfaces because they may fail early. Hard enamel paints lack the flexibility to accommodate the movement of exposed wood. Flexible latex paints are not tough enough to stand up to foot traffic.

In contrast to paint, which flakes and peels, stains "erode" or wear away gradually. Stains must be reapplied more often than paint, but it's an easier job because there is no need for extensive scraping and sanding. Also, weathering stain is less of an eyesore than failing paint.

Sheltered porch floors can be painted with porch and deck enamel. First, treat the deck with a water-repellent preservative (check the label to make sure the product contains a preservative and is paintable). Second, prime the wood with enamel diluted with paint thinner. Last, apply two topcoats of straight enamel. Railings, whether exposed or sheltered, can be painted with latex paints.

Paints and solid-color stains will perform well on pressure-treated wood that is used in an upright position (on fences, for instance), but only when the wood has been cleaned and is thoroughly dry before painting.

Remember, wood to be painted should be dry but not exposed to more than a few weeks of sunlight. The sun's ultraviolet rays damage wood fibers and weaken the wood's ability to hold paint. — *Mark Knaebe*

Painting Cedar Shutters

Q. What's the best way to prep and paint new cedar shutters?

A. Cedar — and red cedar in particular — has the reputation of being a wood that doesn't hold paint well. When red cedar is milled, the wood fibers at the surface get burnished and can form a resinous "mill glaze" that resists paint penetration.

To remove it, use a garden sprayer to spray on a deck wash, then hose off the residue with fresh water. Be sure to clean both sides of each shutter, because if paint starts peeling on the back, it will eventually work its way to the front. After the shutters have thoroughly dried, prime them with a high-quality latex primer.

Next, if you're top-coating in a light color, spray a very light fog coat of a stain-blocking alkyd primer on the front of the shutters. This extra step prevents any remaining tannins in the cedar (which are water-activated) from bleeding through light-colored paints. Then finish up with a top-quality, 100% acrylic latex paint (I recommend Sherwin-Williams's Duration; www.sherwin-williams.com), probably in a semigloss to make the shutters really pop.

If you have a lot of shutters to paint, three coats may seem like an excessive amount of work. But with an airless sprayer (available at most rental stores), you can spray three coats on 50 shutters in a day. In any case, you definitely want to spray — rather than brush on — the alkyd primer, because if you apply anything more than a slight fog, the brittleness will lead to cracking and create problems that far outweigh any benefits. If spray equipment isn't available, buy a few rattle cans of the alkyd primer and just lightly dust the shutters for this step, even if you're brushing the other coats. With modern paints, if you prep and paint right the first time, you'll never have to do it again.   — *Jon Tobey*

Painting Galvanized Steel

Q. I build deck railings, awning frames, and other structures out of welded, galvanized steel tubing. I can't get paint to stick for very long. So far I have only tried Rustoleum after using a mild acid rinse. It lasted about two years before the paint started to peel. Is there a product or technique that will last longer?

A. Yes! Following proper preparation and application of the right product under the right conditions, the paint should never peel, but should provide many years of service before it simply wears away or oxidizes enough to warrant redoing.

We wipe down all new galvanized metal with a rag or sponge saturated with pure white vinegar (that's your acid rinse), then we rinse thoroughly with clear, clean water and let it dry. This is to remove slick manufacturing residues that will prevent your paint product from adhering to the metal. You have to treat all surfaces, because you will have problems with any areas you miss.

Under warm, dry conditions (this is important!), there are many paint products that can now be applied directly to the metal, including most high-end latex paints. Often on gutters and downspouts, we will use our latex or oil trim color as a first and second coat. On railings and awning frames we recommend Benjamin Moore's IronClad Retardo (www.benjaminmoore.com) as both a primer and finish paint because of its unusual durability and its soft, low-luster finish. Whether brushing or spraying, make sure you cover it with a nice full coat on each application. You might be surprised how long it lasts.   — *James Benney*

Removing Stains From Wood Siding

Q. Is there any way to remove mildew and water stains on redwood and cedar siding?

A. To remove a mild case of mildew, scrub the surface with a mild cleanser or non-ammonia detergent. Then rinse with household liquid bleach to kill the mildew spores. Finally, rinse with water.

For more severe mildew infestations, scrub the wood with a stiff bristle brush, using a solution of one cup trisodium phosphate, one cup liquid bleach, and one gallon of warm water. (Wear rubber gloves!) After scrubbing, rinse with a solution of four ounces oxalic acid crystals dissolved in one gallon of warm water in a nonmetallic container. Apply with a soft brush to one entire board or defined area at a time. After the wood has dried thoroughly, rinse with clean water. This will remove not only mildew, but also any water stains and extractive bleeding. It might reduce nail stains as well, however, but nothing short of sanding will remove nail stains entirely.   — *Charlie Jourdain*

Extractive Bleeding on Cedar Siding

Q. I recently installed 20,000 square feet of 1/2x4-inch cedar siding, rough side out, on a project in Connecticut. The exterior wall construction was 6-inch metal steel studs sheathed with 5/8-inch exterior plywood and covered with building wrap. Inside, there is 6-inch batt insulation, a 6-mil poly vapor barrier, and 5/8-inch drywall. The siding was prefinished with a clear sealer at the factory. We are now experiencing "extractive bleeding" in over 20% of the siding's surface area. The supplier and manufacturers of the sealer and siding claim that this is a natural occurrence due to excess moisture. The owner wants the stains removed.

What should I have done to prevent this from happening? Would putting an air space behind the siding have helped?

A. The heartwood of western red cedar (and other species like redwood and cypress) contains dark-colored water-soluble chemical extractives. When the heartwood gets wet, these extractives dissolve, and the solution can run onto the surface of the siding. When the water evaporates, the face of the siding is left stained with brown streaks — extractive bleed.

To avoid this, you have to prevent moisture from reaching the heartwood. The best way to minimize the effect is to pretreat all surfaces of the siding with a water repellent before installation, then treat the face with a semitransparent stain. It's a common mistake to skip the back-priming step. Water gets driven behind siding by wind and is drawn behind siding by capillary suction. Joints, overlaps, and penetrations provide pathways for water to the back of the siding. When the unprotected back gets soaked, extractives can bleed onto the face of the siding below. If you've back-primed, the repellent will shed the water before it can soak the siding and bring the extractives to the surface. Not all clear sealers are good water repellents. Check that the product you used is in fact working to repel water. Spray water onto the surface to see if it beads up.

In general, I think using a clear product on siding is a bad idea. Even the best treatments lose potency through UV degradation in less than a year. Pigmented products are much more durable.

Condensation from excessive interior water vapor can also cause extractive bleed. You have installed a vapor barrier, but air leakage can transport high levels of water vapor into wall cavities. You might consider air-sealing improvements as an option. Providing a vented space behind siding is a good overall rain-management strategy and can minimize the amount of moisture that reaches the back of the siding. It also allows siding to dry more easily.

It's a bit late for that in the project you've already built. You might try cleaning: Extractive stains often come off with a mild detergent if you clean the wood soon after the problem develops. Consider applying a high-quality semitransparent stain like TWP, made by Amteco Products (800/297-7325, www.mfgsealants. com) or Ready Seal (972/434-2028, www.readyseal.com) to your walls once they're clean. These products have fine pigments that help overall durability.   — *Paul Fisette*

Removing Latex Paint

Q. I have trouble removing old latex paint. It doesn't scrape well and it gums up sandpaper. What's the best way to do this?

A. I'm not sure it's the "best" way, but here's what works for me. I use a Porter-Cable disc sander (#7402) that comes with a paint remover attachment. The pad takes 6-inch tungsten carbide discs. I typically use 36-grit, which is pretty coarse and requires a light touch. The problem with a finer grit is that it gums up too quickly.

Often, I'm cutting through six or eight layers of paint. Even with 36-grit, it doesn't take too long for the disc to gum up with latex paint. At around $6 per disc, I can't afford to throw them away, so I remove the discs and soak them in a can of floor stripper — the stuff you use to take up old mastic from vinyl floors. I let them sit overnight, and then have a helper remove the crud using an angle grinder fitted with a wire-brush cup. We lay the grinder on its back, C-clamped to a sawhorse or the tailgate of a pickup. My helper holds the discs with a large pair of pliers and eases them down onto the spinning brush until they're clean. (This task requires rubber gloves and safety goggles.) I typically have 40 or 50 discs on a job.

After removing the paint, I clean the surface with a 3,500-psi pressure washer. This raises the grain, so I come back a few days later and sand by hand or with a vibrating sander. — *Mike Shannahan*

Stripping Paint From Fiberglass Doors

Q. What is the best way to remove paint or varnish from a fiberglass door?

A. Before removing the door from its hinges, wash the door with warm water and common household detergent. Rinse the door well, and allow it to dry. Remove the door to a cool area (55°F to 65°F) and lay it across sawhorses at a comfortable working height. Be sure to cover the floor area to avoid damage from spilled paint remover. Remove the door hardware.

The best paint removers are citrus-based, and can be bought at any paint store. They're effective for removing either latex or oil-based paint and won't damage the fiberglass.

Apply paint remover to a small area (4x4 inches) to test the time needed for the paint remover to work. Depending on the number of layers of existing paint, it can take from five minutes to an hour for the paint remover to work.

When the test is complete, brush a liberal amount of paint remover on the door, brushing in one direction. Remove the residue with a small nylon scrub brush, which will help remove any paint remover remaining in the fiberglass grain. Rinse well, according to the paint remover manufacturer's instructions. — *Ed Fillbach*

Insulating Around Replacement Windows

Q. When replacing old double-hung windows with vinyl replacement units, what's the best way to insulate the counterweight cavity?

A. In our experience, the most effective way to insulate the counterweight cavity is to use blown cellulose. After the stops, sashes, and pulleys have been removed, we tape over the pulley hole, then drill a single 1-inch-diameter hole in the center of the jamb (where it will be covered by the new window). We then pump in cellulose until it backs up and we know the cavity is full.

The advantage of cellulose is that we can pack the cavity tight without risking damage from expansion, as with expanding spray foam. The disadvantage is that we have to lug a blower machine to the job. Fortunately, we have a small one-bag unit, which is relatively portable. These machines can be rented or borrowed from home centers if you buy enough insulation.

If we couldn't put our hands on a blower, our second choice would be to fill the cavity with fiberglass batt insulation. We do this by removing the counterweight access door and pushing small clumps of insulation upward, and by forcing even smaller pieces downward through the pulley hole. We pack the insulation cavity tight to reduce air infiltration.

Until now, we've avoided using spray foams because of the unpredictable expansion rate: It's almost impossible to completely fill a blind cavity without overfilling and bowing out the walls. We were also concerned that, over time, the foam would shrink and leave gaps. I've been told recently that both Great Stuff (Flexible Products Co., 800/800-3626; www.dow.com) and Handi-Foam (Fomo Products, 800/321-5585; www.fomo.com) now have two-part "slow-rise" formulas that are much easier to keep under control. A representative from Fomo Products even told me that if I knew the volume of the cavity, I could call their 800 number and they would tell me exactly how long to squeeze the trigger to dispense the correct amount of foam. I'll probably try it one of these days, on my own house first.

— John Curran

Flashing an Arched Window

Q. What's the best way to flash an arch-topped window?

A. The slickest way I've found to flash an arch-topped window is to use DuPont FlexWrap, a flexible flashing tape from the makers of Tyvek housewrap. I refer to FlexWrap as "peel-and-stick on steroids." Like generic peel-and-stick, FlexWrap is self-sealing (it seals around fastener penetrations), and it has a very aggressive butyl adhesive backing. It costs about $2 a lineal foot. FlexWrap stretches over two times its original length (no, that doesn't mean you have to buy only half as much) and will easily conform to any radius arch-topped window. I also use FlexWrap for the windowsill flashing — the flexibility allows me to fan out the corners and make a one-piece sill flashing.

The sequence of installation is critical in a properly flashed window. First, the window opening is cut out following an inverted "Y" pattern, and a head flap is cut and taped up and out of the way above the window. Next, the one-piece FlexWrap sill flashing tape is installed, followed by the window unit installed in a bead of elastomeric latex caulk. Do not caulk the sill flange — that would trap inside the opening any moisture that finds its way onto the rough sill. An uncaulked sill flange provides a weep area for moisture to escape. Even if the sill is level, shimming the unit provides free space for the water to exit more easily. A fat bead of caulk at the interior side of the sill creates a dam to force water to the sill weep. Jamb flashing tape is installed next, followed by the FlexWrap head flashing tape. To avoid a reverse lap, the head flashing should adhere directly to the sheathing, not the housewrap. The head flap is then folded over the head flashing tape and taped in place. A second piece of housewrap can be cut and inserted in a slit cut above the head flap. This second piece extends at least 12 inches beyond the head flap cut and is backup protection for the head flap taped seam.   — *Carl Hagstrom*

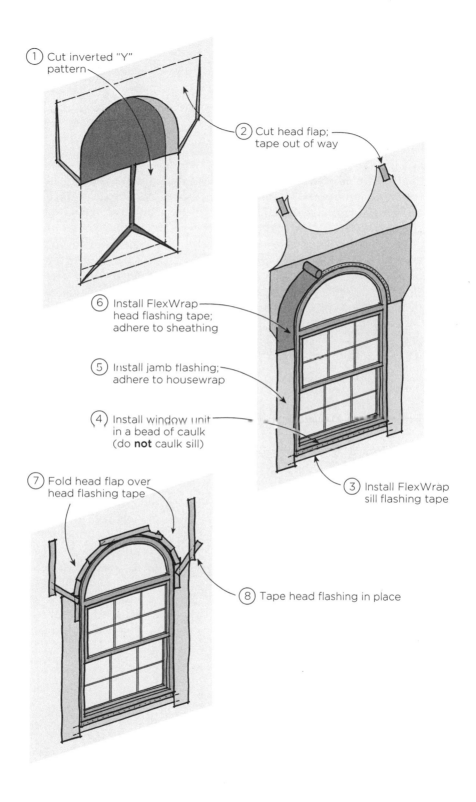

1. Cut inverted "Y" pattern

2. Cut head flap; tape out of way

6. Install FlexWrap head flashing tape; adhere to sheathing

5. Install jamb flashing; adhere to housewrap

4. Install window unit in a bead of caulk (do **not** caulk sill)

3. Install FlexWrap sill flashing tape

7. Fold head flap over head flashing tape

8. Tape head flashing in place

Upstairs Window Egress Rules

Q. I know that the sill of an egress window in a second-floor bedroom must be no higher than a certain distance from the floor to allow for easy escape. But what if it's a tall window that is only 12 inches off the floor? Is this a code violation?

A. All three model codes limit the maximum height of the sill to 44 inches. There is no minimum distance that a window must be from the floor, even for second-story windows. However, if the window extends to within 18 inches of the floor, it must either be safety-glazed or have a bar or other physical barrier to prevent someone from falling into the glazing.

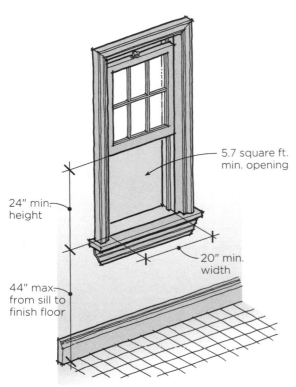

To meet code, the open area of an egress window is measured in "net clear opening area" with a minimum net clear width of 20 inches and a minimum height of 24 inches, plus the opening must be a minimum net clear area of 5.7 square feet. That means a typical 24x20-inch window does not meet code, because it creates only a 3.3-square-foot opening. To comply, the window open area would need to be at least 24x34 inches to equal 5.7 square feet of net clear opening.

— *Kelly Reynolds*

Rebuilding Rotted Windows

Q. As a handyman, one of the problems I see most often is rotted wood trim on windows and doors (for example, brick moldings, sills, and the bottoms of side jambs). Moldings are not usually a difficult fix, but repairing or replacing wood members that are integral to the window or door unit (particularly if it's a double window or a door unit with sidelights) is difficult without weakening the unit. Can you provide guidance about when to repair or replace the rotted wood and how to do it, or whether to simply replace the unit?

A. It's an excellent question. Since specifics could fill a book, I'll offer some generalities.

One's budget will guide most decisions. By and large, unit replacement is the less expensive course of action wherever possible. It's a straightforward job, so the labor costs are usually reasonable. That said, rebuilding is sometimes the only option, especially when you're dealing with historic, discontinued, or custom-fabricated members. A well-equipped woodworking shop should be able to duplicate just about anything. My working philosophy is that if something can be built once, it can be built again.

To avoid future rot, I mill window and door parts from well-seasoned .40 CCA-treated yellow pine. As long as it's dry, CCA-treated wood takes paint well. I leave the use of consolidants, epoxies, and fillers to others. They have their place, but my experience with them has not been overly positive. For milling, a high-quality shaper and bench-mounted production router are musts. There are almost unlimited numbers of over-the-counter bits available, or they can be custom made by specialty tool manufacturers. I buy cutters from Southeast Tool Company in Conover, N.C. (877/465-7012, www.southeasttool.com). Typically, I'll charge the customer for any custom tooling needed and bill on a time-and-materials basis. Since setup can be tedious, I try to mill parts for all the units at once. Your investment in time, skill, and equipment to successfully execute this kind of work is significant, so don't be timid when billing. This is not work for the "we'll beat any price" crowd.

— *Mike Shannahan*

Making a Balcony Door Watertight

Q. I built a house with a second-floor balcony, and I installed waterproofing membrane and tile on the balcony deck. Around the perimeter of the door, I installed Moistop flexible flashing (Fortifiber, 800/773-4777; www.fortifiber.com). The siding is stucco. The room under the balcony is now leaking. Can you provide details on the best way to flash the intersection of the deck waterproofing and door threshold, as well as the best way to flash a door in a stucco house?

A. There are many available deck waterproofing systems, and they aren't all installed the same way, so check with the manufacturer for the correct installation technique. Generally speaking, most leaks occur at termination points, including the transitions between deck and wall metal, deck and door threshold, and deck and door jambs. Other trouble areas include the attachment point for handrails and the area around scuppers and drains.

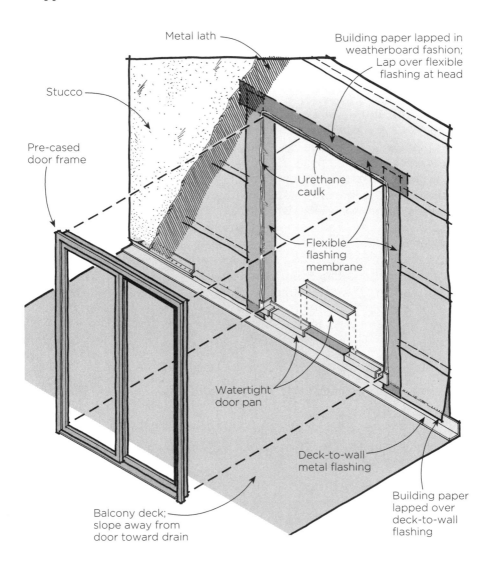

Metal lath

Building paper lapped in weatherboard fashion; Lap over flexible flashing at head

Stucco

Pre-cased door frame

Urethane caulk

Flexible flashing membrane

Watertight door pan

Deck-to-wall metal flashing

Balcony deck; slope away from door toward drain

Building paper lapped over deck-to-wall flashing

All penetrations and terminations should be flashed in a weather-board (overlapping) fashion, with a minimum horizontal overlap of 2 inches and a minimum vertical overlap of 6 inches. Deck-to-wall joints must be lapped and caulked or soldered.

A door will require the following installation details:

1. Install a watertight door pan, like the Jamsill Guard (Jamsill, 800/526-7455; www.jamsill.com). The door pan and finished floor inside must be at least $3/4$ inch above the finished deck.

2. Tack the flexible flashing membrane, such as Moistop from Fortifiber (800/773-4777, www.fortifiber.com), around the perimeter of the rough opening, being careful to lap the flashing over the deck-to-wall metal. The top piece of flexible flashing should overlap the side pieces.

3. Run a liberal amount of good-quality urethane caulk around the rough opening, on top of the flexible flashing. Then install the pre-cased door jamb, pushing the stucco mold trim into the wet caulk.

4. When lathing, be sure to lap the felt or kraft-paper weather barrier in a weatherboard fashion. The felt should lap over the flexible flashing at the head of the door, and over the deck-to-wall metal.

5. If there are any holes, rips, or tears in the flashing membrane or weather barrier, repair the holes. The deck should be sloped away from the door, toward the drains. The deck should be watertight before the base coat of stucco goes on. To check for watertight-ness, take a hose and squirt the deck and then the walls, starting at the bottom and working your way up. If anything leaks, locate the holes and repair them.   — *Ron Webber*

Plumbing & Electrical

5

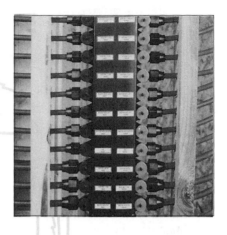

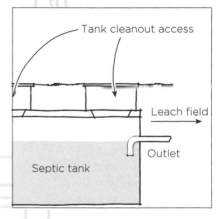

Tank cleanout access

Leach field

Outlet

Septic tank

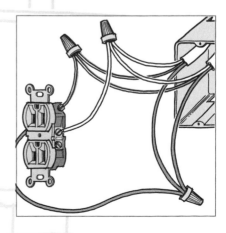

Low Flows & Waste Lines

Q. Do low-flow toilets generate enough water to carry solids through the soil pipe to the city sewage line or the septic system? With so little water being flushed, my concern is that in older homes with cast-iron pipes, corrosion in the pipe may catch the solids and cause blockage.

A. The small volume of flushing water that's available to carry the waste down the pipes is a potential problem, and you are wise to be concerned. With a smooth pipe, such as PVC, there isn't a significant problem until you start using very long runs with large-diameter lines (4 inches or larger). With cast iron, you may have a problem from the onset, because the interior surface is rough and the water does not flow smoothly. Short runs usually present no problem, but long runs can.

You mention older pipes. Actually, if the cast iron has been in for a few years, it has built up a scum on the inside to help smooth out the flow. But there is still no doubt that 1.6 gallons may not be enough water to carry the waste for the entire length of the sewer line by itself.

As you've guessed, there are some tricks we plumbers use to keep ourselves out of trouble. I still do what my Uncle Bud taught me in the 1950s: I always design the system so the waste line of the 1.6-gallon toilet is scoured by another fixture upstream. This way, the additional water from the other fixture will always clean the line — it's just common sense. If this is your situation, you should be okay.

If you have a toilet by itself at the end of a long run of cast iron, you may need a pressure-assist toilet to increase the velocity of the flush. In some cases, the pressurized tank will still not be enough to solve the problem. For example, in my own house, I have a run of more than 250 feet of PVC to the septic tank — 1.6 gallons, pressure-assist or not, simply won't work. For such "big problem" cases, plumbers have been known to modify the toilet to increase the water flow. I never buy a 1.6 that can't be modified. Talk to your plumber about this if you think you've got a major problem situation.

— *Rex Cauldwell*

Venting a Basement Toilet

Q. I am installing an in-floor ejector pump for a basement toilet. Obviously, a wet vent is not possible, since I don't want any upstairs fixtures discharging into the sump pit. How do I vent the toilet?

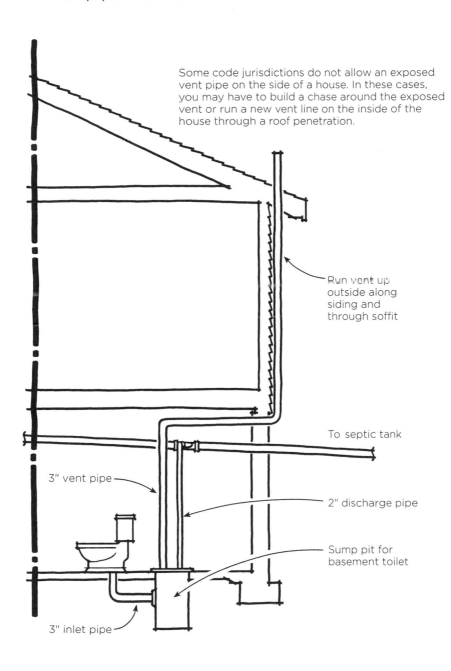

Some code jurisdictions do not allow an exposed vent pipe on the side of a house. In these cases, you may have to build a chase around the exposed vent or run a new vent line on the inside of the house through a roof penetration.

Run vent up outside along siding and through soffit

To septic tank

2" discharge pipe

Sump pit for basement toilet

3" vent pipe

3" inlet pipe

A. You will have two pipes at the top of the sump pit: a high-pressure discharge pipe, usually 2 inches in diameter, and a dry vent pipe, usually 3 to 4 inches in diameter. To connect the vent, first try to find a dry vent in the house plumbing and tap into that. If you have none, you take the vent outside and up along the siding. I've also been known to take it through the floor to an upstairs closet, and then outside and up.

Incidentally, basement ejector pumps do not have to be installed in the concrete floor. Kits are available which allow an above-floor installation.

— *Rex Cauldwell*

Plumbing Around Floor Trusses

Q. We're installing a fiberglass tub unit over a truss-framed floor. A small plan change moved the tub so that the overflow pipe lines up with a floor truss below. There's just enough room to run the drain above the truss, but is it okay to run the drain horizontally for a foot or so beyond where the overflow standpipe connects before putting in the trap? The code we use says there can be no more than 24 inches between the drain and the trap weir, and to place the trap "as close as possible" to the drain, but what does this mean in terms of horizontal distance?

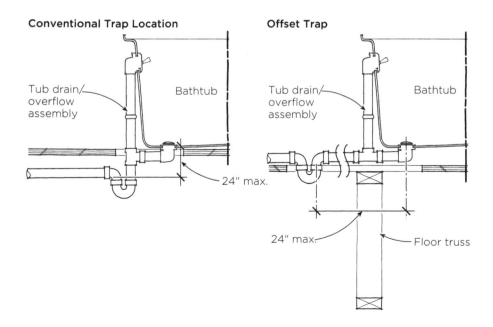

Conventional Trap Location

Tub drain/overflow assembly

Bathtub

24" max.

Offset Trap

Tub drain/overflow assembly

Bathtub

24" max.

Floor truss

A. Most code inspectors will allow up to 24 inches of horizontal distance before the trap where necessary, but be sure to check with your local official. Also, there are offset tub-drain overflow assemblies with side or rear outlets that may help overcome framing obstacles. — *Mike Casey*

Mechanical Plumbing Vents

Q. I saw a "rubber diaphragm" device marketed as a cap for vent pipes that terminated in the attic. Its advantage was that it did not pierce the roof (highly desirable for expensive roofs like slate and tile), yet it would work to equalize pressures in the vent system. No mention of noxious gases or additional ventilation requirements was mentioned. Do these devices work, and are they accepted by the most common plumbing codes?

A. The device you refer to is called an air admittance valve, and was invented by Sture Ericson in Sweden more than 25 years ago. Since then, millions have been sold worldwide by the manufacturer, Studor. According to Studor, the air admittance valve can effectively replace a through-the-roof vent pipe.

Here's how it works. The valve keeps the system closed until it senses negative pressure within the drain system, such as behind a pipe filled with running water. The pressure pulls the rubber diaphragm down as long as the system needs air, and then a spring reseats the diaphragm.

The Studor vent can be installed out of sight, beneath a counter or in a wall or attic. However, most inspectors I know will red-flag an in-the-wall installation. Some will not allow the device at all, so always check with your local inspector before installing one.

I commonly use Studor valves when it is either very expensive or downright impossible to run a daylight vent pipe. I'll also use them for supplementary air in a kitchen or in an existing house that needs more vent air. Although Studor advertises that the system can completely replace the through-the-roof venting system, I prefer to have at least one large-diameter air vent to the outside to provide a way out for positive pressure that might build up in the lines. — *Rex Cauldwell*

Cellular PVC Pipe

Q. What is cellular-core PVC pipe? Is it as strong as solid PVC pipe?

A. Unlike solid PVC pipe, cellular-core PVC pipe is a co-extruded product with at least three different layers. The inside and outside walls are solid PVC, while the inner core is cellular (or foamed) PVC, a material that includes tiny bubbles of entrained air. Cellular-core PVC pipe is available in various wall thicknesses, including Schedule 40 pipe for DWV and thin-wall sewer-grade pipe. For pipe manufacturers, the main advantage of cellular-core PVC is lower cost, since it requires less resin to make than solid pipe.

    The model plumbing codes permit the use of either solid-wall PVC or cellular-core PVC pipe in residential plumbing systems. However, cellular-core PVC pipe is not as stiff as solid PVC pipe. At 5% deflection, a 4-inch cellular-core PVC pipe has a minimum pipe stiffness of 200 pounds-force per square inch, while a 4-inch solid-wall PVC pipe has a minimum pipe stiffness of 310 pounds-force per square inch.   — *Julius Ballanco*

Copper vs. Plastic Plumbing

Q. I'm currently remodeling a vacation home where the copper plumbing has developed pinhole leaks because of an acid water condition. To save money, my client wants to replace the copper pipes with CPVC plastic, but my plumber says that's a bad idea. Who's right?

A. I agree with your plumber. Plastic pipe resists acidic water better than copper does, but it's a "band-aid" solution — the root of your problem is the water, not the pipes. Even if all of your pipes were plastic, the acidic water would cause problems with other parts in the

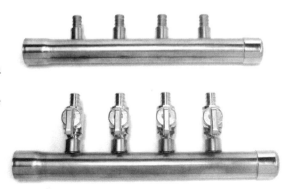

system — riser tubes in the toilet, brass faucets, parts of the water heater, and so on. Also, acidic water affects the flavor of food and drinks made with it. So I advise installing an acid neutralizer.

Acidic water aside, plastic pipe is still an option worth considering. But I would recommend cross-linked polyethylene tubing, not CPVC. The main reason to use plastic tubing in a remodeling job is to save labor, because it's flexible enough to snake around obstacles. It usually requires less demolition, less drilling, and fewer joints.

If you do install flexible plastic tubing, go with a "home-run" system: Use a dedicated line for each fixture, running back to a central manifold panel (see photograph).

A home-run system can use either cross-linked polyethylene, CPVC, or copper. Personally, I'd stick with copper. It's the only product with a proven 50-year track record of success. If you solve the acidic-water problem at its root — which you should anyway — copper plumbing will give you a lifetime of satisfaction, and then some.   — Al King

Polyethylene Water Service Pipe

Q. I would like to use black polyethylene water service pipe for plumbing in the crawlspace of a house that has a black poly service line. Are there any code limitations on using polyethylene pipe?

A. Polyethylene pipe is not approved for in-house cold or hot water lines. According to the International Building Code, once the polyethylene service pipe enters the building, you must terminate it within 5 feet. Whether you consider the crawlspace part of the building is up to you and the inspector.

The intent of the code is to prevent the poly pipe from being used for the in-house water lines. If you use polyethylene pipe to go directly to a water pressure tank without any take-offs to fixtures — or, for city

water, to go straight to a main valve in the crawlspace — you are probably following the intent of the code, even if you have more than 5 feet of
polyethylene pipe in the crawlspace. What you definitely do not want to
do is tee off from the polyethylene pipe. No tees should be installed until
after the polyethylene hits the main valve and transitions to the in-house
piping (for example, to copper or PEX).   — *Rex Cauldwell*

Preventing Water Hammer

Q. What causes hot-water pipes to "bang," and how
can this be prevented?

A. The usual cause of the bang-
ing in domestic hot-water pipes
is water hammer. When the pipe
full of water is running, there is a
considerable momentum from
the mass of water in motion.
When the flow is suddenly
stopped, such as when a solenoid
valve on a washer or dishwasher
closes, or when a faucet is turned
off quickly, this energy of the
moving water is transferred to the
pipe, causing it to vibrate and
move enough to hit the framing.

Sioux Chief Mfg. Co.

Plumbers often take off the
supply from a tee and add a short
length of capped pipe above the
tee to serve as a shock absorber.
This works as long as air remains
in the capped pipe. When the air
chamber becomes waterlogged,
the system has to be drained

*To prevent water hammer, install an
arrester near the fixture. It contains a
bellows and spring to absorb some of the
energy of the flowing water but cannot
become waterlogged the way an
expansion chamber will.*

down to replenish the air. Since air dissolves readily in hot water, however, you may have to drain the system frequently.

The best solution is to install a water hammer arrester (available
from plumbing supply houses under a variety of trade names) near the
fixture. It contains a bellows and spring to absorb some of the energy of
the flowing water.   — *Henry Spies*

Getting Rid of Vent Gas Smell

Q. Is there any practical way to get rid of the smell of vent gas coming up through the traps in a house? The house has a septic tank, and the venting appears to be to code. The problem is persistent, especially after rain.

A. You can try replacing the closet gaskets, but that may not fix the problem. Also check for missing traps at all the sink, tub, and shower locations. If nothing is amiss, there is a solution we have used successfully many times over the years.

Septic tank systems are missing one element that we're required to install on sanitary sewer lines connected to municipal systems — namely, a trap set in the main line. When dealing with the problem you describe, we install a standard 4-inch trap set consisting of a sanitary tee laid on its back, a house trap, and two cleanouts (see illustration).

The trap set should be installed with a 1/4-inch-per-foot fall and laid in crushed stone to prevent it from settling back and lying in the wrong direction. The trap stops the septic tank odors, while the fresh-air vent permits free passage of air by natural convection up and out through the roof vents, carrying away any foul odors within the home's drainage system.

The first time I did this, I worried a bit about creating an air-bound situation, given that the tank is unvented. But, in fact, what goes into a septic tank comes out the other side, and its air volume doesn't change sufficiently during that process to create any problems. So the technique solves the problem and improves the home's venting system.

It may be tempting to combine the mushroom vent with the house-trap cleanout, but don't do it: Technically, that would create what's called a "crown vent," which is disallowed under most plumbing codes. The reasoning behind this is the fear that the trap seal might be lost through evaporation as air is drawn up through the mushroom vent.

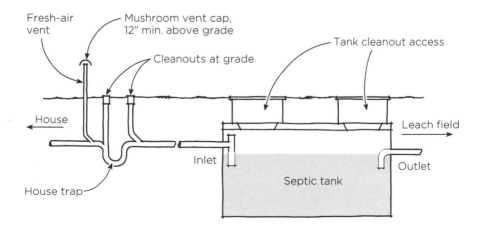

Fresh-air vent — Mushroom vent cap, 12" min. above grade — Cleanouts at grade — Tank cleanout access — House — Leach field — Inlet — Outlet — House trap — Septic tank

By the way, we rarely see clogs in house traps unless the occupants have flushed things down the line that had no business being there in the first place, such as grease, paper towels, rags, pencils, pens, false teeth, and cups! I thought the advent of 1.6-gallon-per-flush water closets would mean constant trap clogs, but I have not seen any additional stoppages created by their use.   — *Dave Yates*

Reliability of Low-Flow Toilets

Q. My remodeling customers are frequently skeptical when I replace their old 5-gallon toilets with 1.6-gallon toilets. They have heard of problems like "skid marks," multiple flushes, and so forth. Also, the word I have heard from the field is that pressure-assisted toilets create more problems than they are worth. Have the new low-flow designs solved these problems?

A. When 1.6-gallon toilets first came out, they did not work well because they were not designed for low flush: They were essentially 3.5-gallon toilets with a 1.6-gallon tank capacity. Everyone had trouble with them (you had to flush twice), so many consumers resorted to pressure-assisted units. Pressure-assisted units typically work well in the beginning (although many are noisy), but may eventually need repairs. In some cases, parts may be hard to find.

I prefer to keep it simple and use a gravity toilet. Most of the new 1.6-gallon designs work fine. However, when replacing a 3.5-gallon unit with a low-flow, I try to keep the customer's expectations in line. "Skid marks" are a problem with all the units and are just something they'll have to live with. Some manufacturers give a written one-flush warranty and advertise quiet operation, but I make no personal guarantees.

To make the low-flows work better, many manufacturers reduced the toilet's passageway diameter to around $1^1/2$ inches. This gives the water the velocity it needs to evacuate the bowl, but the narrower outlets tend to clog more often. One manufacturer, Toto, advertises a much larger passageway — $2^1/8$ inches — and other makers are following suit.

A larger passageway — or a higher price tag, for that matter — doesn't necessarily mean the toilet will flush better. Some manufacturers have designs that are just downright stupid. One well-known unit has to have the handle in the down position for the entire length of the flush. You can't just push the handle down, let go, and assume it will flush. If I'm asked to install this unit, I'll make minor modifications to get it to work better. In general, low-flow toilets are more sensitive to minor problems than 3.5-gallon units. For example, putting blue tablets in the tank causes a sticky coating to form everywhere. Older toilets can tolerate this, but with low-flows, it can slow down the water, causing "lazy flush syndrome." And even a minor misadjustment of the water level refill system in a low-flow unit may only allow 1 gallon to enter the tank — not enough for a successful flush. By contrast, if a

3.5-gallon toilet is misadjusted so that the tank only has 2.5 gallons, the toilet will probably still flush fine. Ask your plumber for advice when you're selecting a low-flow toilet. He or she should know which models give the best service.

— *Rex Cauldwell*

Setting a Toilet Over Tile

Q. What is the proper technique for setting a toilet on an irregularly tiled floor?

A. Most floors are a little uneven — with tile floors being the worst — so toilets often need to be shimmed when they're set. I like to set the toilet down first without wax to get a better idea of where the high and low spots are. Then, when I set the toilet, I use either a wax seal, which works well with older cast-iron and lead DWV systems, or the Fluidmaster Wax-Free Bowl Gasket (www.fluidmaster.com), which seems to work best with newer plastic waste lines.

To shim the toilet, I generally use cedar door shims because they have a more gradual taper than the plastic ones I've seen. I try to shim from back to front so that the front of the bowl is touching the floor, and from side to side so that the toilet is level. Shims usually end up at the back of the bowl, where they're held in place by the weight of the toilet.

After the toilet bolts have been snugged up, I caulk at the front of the bowl, from bolt to bolt, for sanitary purposes. If a leak develops only around the gasket, it's better to know about it than to conceal it behind a continuous bead of caulk, so I leave off the caulk around the back. — *Terry Love*

Shower Pipe Knocking

Q. I am having a problem with a shower valve in a bathroom remodeling project. When you turn the water all the way to hot and then back off to around the middle of the temperature range, you get a pulsating water flow from the showerhead and severe, loud knocking from the shower pipes in the wall. The problem does not occur when the valve is opened all the way, full hot. We've secured the valve and the piping with clamps. Are you aware of any remedy for this problem short of taking the tiled walls apart to install a water hammer arrester?

A. A water hammer arrester won't solve the problem; I think the problem is in the faucet body itself. This type of water pulsation and the associated water hammer are typical of faucets with balancing valve assemblies and check valves. The problem is not limited to one manufacturer; I've experienced it with several brands.

You have a choice of pulling the valve cartridge and checking it for debris or just replacing the faucet. Most of the time, I replace the faucet, because, even with new parts, I often can't fix the problem. I typically switch the faucet to a Moentrol, by Moen; I've had the best luck with that brand. — *Rex Cauldwell*

Buildup in Shower Drains

Q. Recently I've run into a couple of incidents where a crystallized white substance has formed on the shower floor and in the drain. In both cases the showers have mud-set floors and are entirely tiled. The water is from a public source. In one case, the drain was almost entirely blocked. To clean it, we had to take a screwdriver and chip it away. Is this a chemical reaction of the tile grout with cleaning or shampoo products?

A. There are a couple of possible culprits. The first is efflorescence, which occurs early in the life of a mortar bed installation when minerals from the sand-cement-lime mix get deposited on the tiles. It usually happens because too much of a particular ingredient — hydrated lime or Portland cement, for example — has been used. The shower water brings salts and minerals to the surface of the tile, and a white deposit is left behind when the water evaporates. If materials within the mortar bed, adhesive mortar, or grout are the problem, the efflorescence should go away after 28 days — the curing period for Portland cement products.

Efflorescence can also be caused by salts or minerals being carried by an outside source of water. This happens frequently when ground water seeps through a foundation wall or slab. The cure here is to stop water before it can enter a structure.

The buildup you're seeing might also indicate that the weep holes in the shower drain are clogged. If the water moving through the mortar bed to the weep holes can't exit, the mortar bed will become saturated with water that will wick upward into the wall-setting bed materials, or through the floor tile grout joints where it evaporates and leaves its mineral cargo behind on the surface of the tiles.

Hard water is another possible explanation for the buildup. You mentioned the water is from a public source, so if the water is hard, it would be common knowledge. Evidence of hard water is easy to find: Look for deposits and crust on showerheads and tub spouts, and for visible water-line marks around the inside of toilet bowls. The best solution is to install a water softener; otherwise, the buildup will continue and may eventually clog the shower drain's weep holes. Cleaners that are strong enough to remove lime, salt, or other mineral deposits and yet safe enough for use with tiles are available from most tile supply stores. — *Michael Byrne*

Overrated Service Panel

Q. Can a 100-amp service panel be used with a 60-amp incoming service?

A. For all practical purposes, no. The service rating for a panel comprises the amp rating of both the incoming feed conductor, the over-current device, and the panel. A service panel main breaker rated at 100 amps would trip only after the service load exceeded 100 amps. Since this is well above the 60-amp capacity of the service wire and the utility meter base, all would be overstressed and not protected.

Installing a 100-amp main lug panel outfitted with a 60-amp main breaker is considered "backfeeding." This is only allowed if the breaker is mechanically installed — with a screw, for example — and not just pushed in. The common breaker in a residential panel doesn't give you that option. — *Rex Cauldwell*

Installing 20-Amp Breakers

Q. In a residential rewiring job, I am removing the existing conductors from steel flex conduit and installing new 12-gauge solid wire, pulled through the flex conduit. Can I convert all of the existing 15-amp breakers to 20-amp breakers?

A. Yes, you can. Since you have installed 12-gauge wire, and since the breakers are there to protect the wiring, there is no reason you can't install 20-amp breakers — as long as you don't put more than three conductors in the conduit. If you install more than three conductors in the conduit, you would have to start to derate the conductors. — *Rex Cauldwell*

Does Panel Location Matter?

Q. In a new house I am building, I would like to locate the electrical service panel under a counter. This would put the top of the panel at a height of about 34 inches. Is there a reason this shouldn't be done?

A. There are more reasons not to do this than I can cover in this limited space. First of all, you cannot put the panel under anything. You cannot put it over anything. You cannot put it next to anything. There must a clear space in front of the panel where an electrician can stand. The space must be kept clear for at least 3 feet in front of the panel, and for 30 inches left to right in front of the panel. There must be at least 6 feet 7 inches of headroom, measured from the floor to the ceiling. The main breaker can be no higher than 6 feet 7 inches above the floor. In your case, locating the panel under a counter would require an electrician to work on his or her knees when working in the panel. For this reason alone, the inspector would probably fail you.

— *Rex Cauldwell*

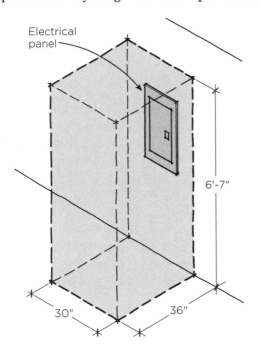

Electrical panel

6'-7"

30"

36"

Should Older Wiring Be Updated?

Q. When remodeling or adding to an older home, what are some things to look for that indicate the entire house's wiring should be updated or replaced?

A. If you see any of the following conditions, you should think seriously about rewiring:
- any system without a ground (bare) wire
- cloth insulation, or any worn insulation that deteriorates in your hands when you move the wires (this leaves bare spots along the wire, creating the potential for dangerous shorts)
- less than 20-amp circuits in the kitchen, laundry, or bathrooms
- lights that dim as a result of an appliance being turned on
- a fuse box, because it generally indicates an obsolete system (chances are good that you need new wiring, a new panel, and probably a grounded, or GFCI-protected, system)
- the presence of branch-circuit aluminum wire (if you find this, call an electrician)

— Cliff Thomas

Must Remodel Work Come Up to Code?

Q. If an electrician comes and does work, does he have to bring the whole house up to code? Are there certain things he has to do and others that are optional? What is permitted versus what is best practice?

A. I think the code is specific enough to make it clear when violations are present. I would ask the local inspector to clarify any concerns or gray areas. Meeting the requirements as set forth by code and inspectors is the only sure way to cover yourself, and liability is a real concern when doing electrical work. I am not aware of any specific National Electrical Code references to old work correction requirements. Typically, the rule for remodeling a residence is that if 50% or more of a building is being affected, then the entire building must be brought up to code.

If you don't know exactly what you're looking at and how it should look, call someone who does. This stuff (electricity) is responsible for deaths every year that can be avoided. *— Cliff Thomas*

Is Romex Getting Smaller?

Q. After my electrician finished rough wiring the last house I built, I noticed that the Romex (NM) wire he used seemed to have a thinner profile, and the individual wires seemed smaller in diameter than what I was used to. Is the copper wire (or the insulation) getting smaller, or is it just my imagination?

A. No, your imagination is not getting the best of you — the nonmetallic sheathed (NM) cable used in most residential construction has become smaller.

NM wire is made up of four basic components: the copper conductor, the conductor insulation, the cable filler, and the outside jacket. The diameters of the different-gauge conductors have remained the same for years and are unlikely to change.

But recent improvements in the plastics used to produce the three remaining components have resulted in a more compact NM cable with better insulating characteristics and a higher abrasive resistance.

The most common conductor insulation today is THHN, which measures 15 mils on a #14 conductor, whereas the older THW conductor insulation measures 30 mils. The cable filler separating the conductor insulation from the outside casing is now made from plastic and is thinner than the traditional paper filler found in older NM wire. The thickness of the outside jacket has also been reduced. The result is a NM cable with a smaller circumference, a tougher outside casing, and a smaller conductor area that requires less filler.

How much smaller is the new NM cable? It used to be that only two 14/3 NM cables would fit through a 1-inch-diameter hole — now you can fit three.

— Eric Lewis

Rules for Splicing Romex

Q. I want to lower receptacle outlets from 4 feet off the floor to 18 inches. I plan to make a splice at the existing receptacle and extend the new wire down through the stud cavity. Can the junction box be inside the wall or does code require that you have access to it by using the existing outlet box as a junction box with a solid cover plate?

A. You'll have to use the exposed cover plate. By code, you must leave access to any splice anywhere. (The only exception is an irreversible crimp on a service entrance cable — a special case requiring an expensive crimping tool.) Splices in attics and crawlspaces are considered accessible as long as there is a hatchway into the space. Even these splices must also be in a box and behind a cover plate. — *Rex Cauldwell*

3/4-Inch Conduit Capacity

Q. What is the maximum number of 12-gauge wires permitted in 3/4-inch conduit?

A. For practical purposes, the maximum number of 12-gauge THHN conductors for a 3/4-inch EMT conduit is nine.

The answer would be different for another type of conduit (for example, ENT) or if the conductor had a different insulation type or different outside diameter. Although the NEC allows up to 16 current-carrying THHN conductors in such a conduit, it also requires that for any number of conductors over three, you have to derate the capacity of the conductor. The derating isn't significant until the number of conductors exceeds nine. For 10 to 20 conductors, 12-gauge wire is derated down to 15 amps. — *Rex Cauldwell*

Tracing 3-Wire Circuits

Q. In many of the houses we work in, we are asked to evaluate the electrical system. We sometimes find 10/3 wire used on 110-volt circuits with 15- and 20-amp breakers in the panel. The installer has used the red and the black wire for separate circuits. Is this an acceptable installation, or does it need to be pulled out and changed?

A. Whether or not this is an acceptable wiring scheme depends on where the wire leads.

Partially switched receptacle. Three-conductor wire has two hots — black and red — and a white neutral. Though normally used for three-way switching, three-conductor wire is commonly used for duplex receptacle wiring as well. For a partially switched receptacle, for example, you would break off the tab on the brass side of the receptacle. The black and red

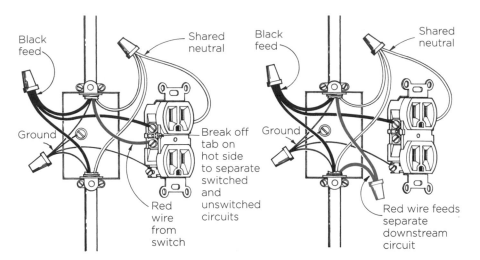

Black feed · Shared neutral · Ground · Break off tab on hot side to separate switched and unswitched circuits · Red wire from switch

Black feed · Shared neutral · Ground · Red wire feeds separate downstream circuit

Illustration A. *To wire a partially switched receptacle, break off the tab on the brass (hot) side of the receptacle. Then connect the black and red wires of the incoming feed to the now independent brass screws. Leave the tab on the neutral side intact. The top of the duplex receptacle is now permanently hot, while the bottom is switched. Both top and bottom share the same neutral.*

Illustration B. *Three-wire Romex can be used to power two separate circuits that share the neutral. For example, here the black wire feeds a receptacle circuit, while the red feeds a lighting circuit. Although multiwire branch circuits are permitted by code in certain cases, they are not recommended for residential wiring.*

wires are then connected to the now independent brass screws, one wire to each; it makes no difference which wire is on which screw. The neutral side of the receptacle is left intact. This makes the top of the duplex receptacle independent from the bottom, while they both share the same neutral (see Illustrration A). The black or the red can go to a permanently hot feed, while the other goes to a wall switch, creating a switched outlet. This layout works well for switching on a table lamp in a room with no overhead lighting, but it doesn't require heavy 10-gauge wire.

If you find a red wire and a black wire from the same three-wire cable coming into the panel, then going to separate single-pole breakers, you'll need to check the installation. This scenario isn't automatically wrong, but to be safe I would assume it was until I traced the wiring back from the panel. It could be one of several different things, but it's usually a multiwire circuit or a 220-volt circuit.

Multiwire circuits. Three-conductor wire can be used to power a single circuit that would otherwise require two 2-wire circuits. For example, the black might feed a line of receptacles, while the red feeds a line of recessed light fixtures in the same area. The white is common to both (Illustration B). If the loads on each wire are nearly balanced, the neutral wire carries only the unbalanced current, resulting in less of a voltage drop and wattage loss than the two 2-wire circuits would have. In this situation, the black and the red must be connected to the opposite phases in the panel to prevent overloading the neutral (white) wire.

This type of wiring is called a multiwire branch circuit. Though common in many areas, I do not recommend it. This system is governed by specific code regulations. A multiwire branch circuit must be able to disconnect both phases (that is, both branches) at one throw of the breaker. Therefore, these circuits must be ganged together. If someone does split the branches apart, and they both lead to the same phase on the panel, the neutral currents would be added together. Such an overload could have disastrous results. In addition, the receptacles, switches, and other devices on multiwire branch circuits must always be pigtailed, as shown in Illustration B. This is done to ensure that the neutral is not interrupted by a poor mechanical connection on the device (a wirenut is a much more reliable connection). Because of these requirements and the problems associated with multiwire circuits, my advice is not to use them for residential wiring where do-it-yourselfers are likely to muck around with them.

220-volt circuits. A similar scheme (red and black to opposite phases of the panel) is often used by do-it-yourselfers to power electric baseboard heaters or some other 220-volt circuit. But do-it-yourselfers often don't know to use a double-pole breaker or don't remember to connect the tabs between the breaker handles. This might also explain the 10-gauge wire: Someone incorrectly assumed the heaters needed the heavier wire. Do-it-yourselfers often don't know that the breaker protects the wiring.

The bottom line is: Trace the wires. See where they're going and what they're connected to. Or throw the breakers and see what doesn't work. Then trace it down.
— *Rex Cauldwell*

Insulating Around Knob-and-Tube

Q. I am remodeling a 1920s house that has the original knob-and-tube wiring. I would like to blow cellulose into the stud cavities but I am guessing that it's not safe to do that with the exposed wiring in the cavities. What does code say?

A. You're right; it isn't safe. The reason is that knob-and-tube wiring is designed for open-air heat dissipation. That means there cannot be insulation around the wires, which would be the case if you blew cellulose into the wall cavities. The same applies to knob-and-tube wiring in the attic or between floor joists. You shouldn't have any insulation around the wiring — not even fiberglass.
— *Rex Cauldwell*

Installing Wiring Between Ceiling Strapping

Q. Here in New England, we usually attach 1x3 strapping to the ceiling joists before installing drywall. At most jobs, I see the electricians stapling the Romex cable to the bottom of the joists, parallel to the 1x3 strapping, so the wiring is about 1/2 inch or less from the back of the drywall. Does this meet the NEC, or does the wiring need to be recessed farther from the drywall?

A. During the 12 years I worked as a residential electrical contractor, I routinely stapled cable between the strapping. All of the inspectors I encountered in Massachusetts, New Hampshire, and Maine accepted the practice.

Unfortunately, though, the National Electrical Code (NEC) is not clear on this issue, and some local inspectors may interpret the NEC differently from the inspectors I encountered. NEC Section 300-4 states, "Where subject to physical damage, conductors shall be adequately protected." In a perfect world, the NEC would address the issue of whether cables buried just under the surface of drywall are subject to physical damage. But we do not live in a perfect world.

NEC Section 300-4(d) may apply, but only if your local inspector considers strapping to be a "framing member." That section states that "where a cable- or raceway-type wiring method is installed parallel to framing members ... the cable or raceway shall be installed and supported so that the

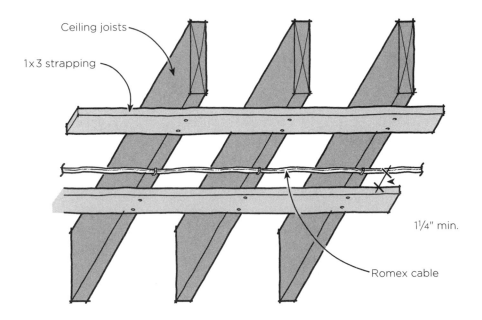

Ceiling joists

1x3 strapping

1¼" min.

Romex cable

nearest outside surface of the cable or raceway is not less than $1^1/4$ inch [es] (31.8 mm) from the nearest edge of the framing member where nails or screws are likely to penetrate." The intent of this section is to keep the cable away from the zone where drywall nails and screws can be expected. This section does not require the cable to be $1^1/4$ inches back from the drywall — only $1^1/4$ inches away from the strapping. — *Sean Kenney*

Wiring in Stress-Skin Panels

Q. How is wiring accommodated in a house built with stress-skin panels? Is it possible to rout a channel in the panel to retrofit a wire?

A. Most manufactured panels have built-in electrical channels. These channels are typically 1-inch holes running through the center of the panel section, so they are well out of the way of drywall screws. They usually run horizontally 12 to 16 inches off the floor, or at about 44 inches (countertop height).

In addition to the horizontal runs, the panels have vertical channels. In some cases, these are placed in the center of the panels, so you have a rise for the wiring every 48 inches. In other cases, there is a notch behind the 2x4 spline that joins each panel. In either case, you have to drill a 1-inch hole through the plates as you set the panels. You then cut in switch and outlet boxes and fish your wires. Obviously, this process is more time-consuming than running wire in a conventionally framed home. And since fishing wire horizontally is so time-consuming and difficult (especially around corners), it's best to run more of the wiring in the floor cavity than you might ordinarily. In slab-on-grade homes, you have to run some wires as you set the panels, which lengthens the process even more.

For that occasional added light switch or outlet, you can rout a channel out, lay the wire in, and then foam it in place. Unless the panel has integral studs, never rout more than about 8 inches horizontally, or you'll destroy the strength of the panels' skin. Vertical runs should stop at least 14 inches from the top and bottom of the panel, which means you have to have a pretty long bit to finish the hole and drill through the plates.

When foaming the wires in, I typically bring the foam flush to the surface of the panel's skin. Later, this foamed-in channel will be covered by drywall. In the case of a retrofit, when the drywall is already up, leave the foam recessed, so you have room for mud to patch over the channel. To avoid a hassle with the code inspector, you have to rout at least a $1^1/4$-inch-deep channel (or, when drywall is up, at least $1^3/4$-inch-deep), so drywall screws will not pierce the wire. The bottom line with all these recommendations is: check with the manufacturer for a choice of options and specific guidelines for using its product. — *Steve Andrews*

Grounding Steel Studs

Q. Are there any special requirements for electrical grounding when using metal studs?

A. Wherever steel studs are used to support NM cable (the plastic-jacketed cable commonly used in residential construction), there is the potential for the framing to become energized. Metal electrical boxes or metal equipment that might be attached to steel studs must always be grounded, which effectively grounds the studs. When plastic device boxes are used, I recommend installing a bonding jumper to the steel frame. This jumper should be sized to match the highest-rated breaker that feeds the specific section of wall. In some areas, the electrical inspector may require a bonding jumper for steel boxes as well.

While the National Electric Code (NEC) does not specifically address the question of grounding a steel frame in residential construction, Section 250 of the NEC contains two citations that mention "structural metal frame of a building" as applied to large steel structures. Section 250-58a says the metal frame must not be used as the equipment grounding conductor for alternating-current equipment. The intent here is to make sure that any electrical faults are directed to the equipment grounding conductor — not through an unreliable path like the steel beams or studs.

Section 250-81b says the metal frame must be part of the Grounding Electrode System (GES) when the frame is intentionally grounded. The GES primarily provides an alternate path for external voltages (lightning and utility surges) so they do not find their way through the structure. In residential construction, steel studs are often used in conjunction with wooden laminated beams or treated wood mudsills, both of which act to isolate the steel frame. But if the entire structure is framed in metal and lightning is a significant threat, then the structure should be effectively bonded to the GES. (For more information on lightning requirements, refer to National Fire Protection Association publication 780-1992 [ANSI] and check with your local jurisdiction.) — *Redwood Kardon*

Copper Pipes and Armored Cable

Q. If a copper water pipe is in contact with the metal sheathing of BX cable, can that contact cause the cable to corrode?

A. Yes. The flexible sheathing on BX or AC (armored cable) is made of aluminum or galvanized steel, and contact with a copper water pipe can cause it to corrode over time.

Galvanic corrosion occurs when dissimilar metals are placed close together in the presence of an electrolyte like water. In this case, condensation on the pipe functions as the electrolyte. The zinc coating or aluminum (anode) is sacrificed to the copper (cathode), which remains intact.

There is perhaps a more important concern than corrosion: Such contact can be dangerous. The metal sheathing on the cable functions as the exclusive grounding path for the wire. In case of a short or ground fault, a pipe touching the cable could be charged and become a hazard.

— *Paul Fisette*

Tying Into Old Armored Cable

Q. Can you keep old armored-cable circuits, or should you replace them? If you disconnect old circuits, what's the rule around tearing out old wire versus leaving it buried?

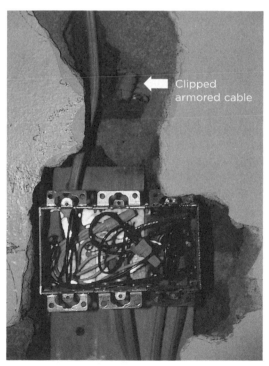

Clipped armored cable

If you abandon existing work, make sure it is cut far enough back so it can't be reconnected, like the armored cable near the top of the photo.

A. I would never add on to an old armored-cable system. A ground wire has been included in armored-cable wire for only the last 10 years or so, which means you could be dealing with an ungrounded system. Better to leave the old system to itself and make all your work new and safe. When you disconnect old circuits, it's important that you make them unavailable for connections in the future (see photograph). Remove exposed sections of wire to avoid future use or confusion when tracing other wires. This will also make for a cleaner job.

— *Cliff Thomas*

Connecting to Aluminum Wiring

Q. I recently did some work in a house where the electrician found aluminum wiring. The electrical inspector told us that pigtailing in standard devices with wire nuts was not acceptable. Instead, we must use aluminum-rated devices and special aluminum-to-copper connectors. I checked with the local supply house, but they said they couldn't get aluminum-rated devices and had never heard of wire nuts rated for aluminum-to-copper connections. Any ideas?

A. Though large-diameter aluminum wiring is common in households, aluminum branch circuit wiring is not. At one time, some houses and mobile homes were wired with aluminum. Problems arose with the devices, which were designed for copper. The expansion and contraction of dissimilar materials caused the connections to loosen, and this sometimes created an arc that would start a fire.

To my knowledge, aluminum branch circuit wiring has since been discontinued, but it's still around in many existing homes. Contractors who encounter it should take certain precautions: Nothing can be connected to aluminum wiring unless the device is labeled CO/ALR. Appropriate switches and receptacles are available from several manufacturers, such as Leviton (718/229-4040, www.levitonproducts.com).

I do not know of any light fixtures that allow aluminum wiring. (The silver you see on some light fixture wires is tinned copper.) If the light fixture needs to be replaced, a new one can be installed using SWS bi-metal-tongue, split-bolt connectors from Penn-Union (814/734-1631, www.penn-union.com). Be very careful with the hookup. Do not bend the aluminum or it will break. Once the connection is tightened down, cover it with an antioxidant, then cover both the connector and 1 inch of the wire with electrical tape. This will support the aluminum wire, as well as insulate the splice. Purple wire connectors are designed for aluminum-to-copper splicing, but they are controversial, as some critics consider them to be fire hazards.   — *Rex Cauldwell*

Switching From 3-Wire to 4-Wire

Q. How do you wire a four-wire 240-volt appliance (kitchen range or dryer) in a location where there used to be a three-wire appliance?

A. You have to replace the cable with a 4-conductor cable with matching receptacle. Here in the Southwest, most houses are built slab-on-grade and have flat roofs. If you don't have an attic or crawlspace, and the appliance is on or near an exterior wall, you can install new conduit on the exterior. Between the exterior wall and the appliance, there are often cabinets behind which you can hide conduit, with little intrusion or patching. Flexible conduit can be helpful here.   — *Cliff Thomas*

Tips for Snaking Wires

Q. I have to rewire an old home that has well-maintained plaster walls throughout. I'd like to do this without gutting the interior. Any tips for wire-snaking?

A. When rewiring an old house, I'll wire the first floor by working up from the cellar (Illustration A), then run a few feeds up to the attic and wire the second floor by working down from there, using drill bit extenders, fish tapes, and a tone generator as needed. It's usually easy to find interior wall top plates in the attic just by moving aside the insulation (Illustration B). By drilling at an angle from inside rooms, you can usually get through top and bottom plates and into stud cavities (Illustrations C & D). When I have to run wires horizontally through studs, I'll try to minimize repair work by creating just one hole at the front edge of each stud (Illustration E).

While doing the job, try to coordinate with other crews working in the house at the same time. If, for instance, the plumber has to open a wall anyway, you may as well run your wires before it's closed up again.
— *Sean Kenney*

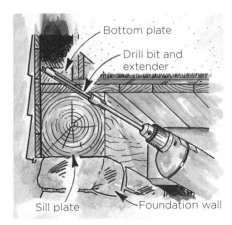

Illustration A. When an old sill is deeper than the wall, use a bit extender and drill at a shallow angle.

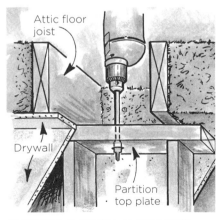

Illustration B. Finding partition walls from the attic is usually easy — just lift the insulation and look for the top plates.

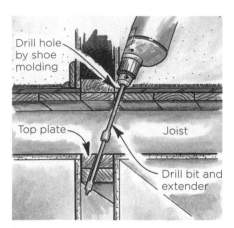

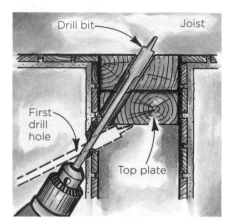

Illustration C. *By removing the shoe molding at the edge of a carpeted room, you can drill an inconspicuous hole through the floor.*

Illustration D. *After drilling one hole to find the bottom edge of the top plate, you can drill at an angle that will pop the bit through the center of the wall in the attic.*

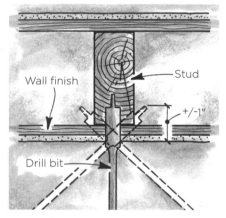

Illustration E. *To run a wire across a stud or joist, drill three times, using the same entry hole: one straight in and one each at a tight angle into each bay.*

Placing Wall Receptacles

Q. An old house I'm working on has too few wall outlets to meet code. What are the rules for spacing and positioning of new outlets? Can I place them in the baseboard trim? And do some of them also have to be switch controlled?

A. Most old houses have too few receptacles to meet current code requirements.

Any wall section 2 feet wide or wider requires a receptacle, and every point along any wall must be within 6 feet of a receptacle. Doors and fireplaces don't count as part of the wall, but fixed-glass panels (like the nonsliding half of a glass slider) do. So starting at a door frame or corner, you must place a receptacle within 6 feet, and one at least every 12 feet thereafter. Receptacles dedicated to one specific appliance, floor receptacles more than 18 inches away from the wall, and receptacles more than $5^1/2$ feet from the floor do not count as required receptacles. This is code minimum.

For a premium job, I suggest going beyond code and adding one receptacle on each wall within 3 feet of a room corner, one on each side of any window 3 feet wide or wider, and one on each side of the bed (assuming the bed never moves).

Receptacle height is not specified as long as you don't exceed $5^1/2$ feet from the floor. Receptacles can be installed above that height, but they are not counted as part of the required minimum.

And yes, receptacle outlets can be placed in the wood trim. I do this quite often in log cabins and renovations. However, be careful if you try to remove outlets from or place them in antique wood trim — the wood is easy to damage and hard to replace.

No receptacle is required to be switch controlled. However, you are required to have switched lighting in most habitable rooms. This is usually done with an overhead light in the ceiling, but a floor lamp plugged into a switched receptacle also satisfies the requirement.   — *Rex Cauldwell*

Upgrading Two-Prong Outlets

Q. Is it possible to replace old two-prong outlets with modern three-prong receptacles?

A. Upgrading outlets from two-prong to three-prong is allowed by code only if the circuit is GFCI protected. If you do add grounding-type receptacles to an ungrounded circuit, you must not wrap any ground wire around the grounding terminal on the new receptacle. Any GFCI without an attached ground wire must be marked "No equipment ground." Also, any GFCI-protected outlet without an attached ground wire must be marked "GFCI protected. No equipment ground."

To upgrade a two-wire system, you can either replace all of the breakers in the panel with GFCI breakers, or place a GFCI feed-through outlet at each home run location. Although breakers are expensive, locating the home run outlet in each circuit can be time consuming. Using breakers is a labor-saving and reliable approach.   — *Cliff Thomas*

Push-In Connections on Receptacles

Q. Many of the receptacles I use have push-in connectors on the back. Does a push-in connection perform as well as the side-mounted screw terminal connection?

A. Sometimes I get residential service calls where the power is out in a receptacle or a portion of a room. About 75% of the time, the problem is caused by a failed push-in terminal on a receptacle or switch.

Push-in terminals are small copper spring clips that only make contact with a small portion of the wire. When a push-in terminal is subjected to a high-amperage draw, the terminal often overheats, eventually causing the connection to burn out. Another problem is that when the electrician pushes the receptacle or switch into the box, the wires twist and bend, putting a lot of stress on the relatively weak spring clips.

Even though the push-in connectors are quicker and easier to use than the screw terminals, I never use them, nor do I allow my employees to use them. Note, too, that the National Electric Code (NEC) has restricted the use of push-in terminals to #14 AWG copper wire only.

Some ground-fault circuit interrupter (GFCI) receptacles have clamp-style terminals that are almost as easy to use as push-in terminals. The wire is pushed into the back of the GFCI and a screw is tightened to secure the clamp. Many higher-grade receptacles and switches have a similar clamp terminal. These terminals make a good connection, but as with any terminal, the wire should be given a tug to make sure the connection is tight.   — *Sean Kenney*

Receptacles on Ungrounded Electrical Circuits

Q. My questions have to do with existing ungrounded circuits in older homes. My understanding is that you should always replace worn-out two-slot receptacles with new two-slot receptacles (although they're not always easy to find). However, I've also been told that in some locales you can install a three-prong receptacle in an ungrounded circuit as long as you fill the ground slot with epoxy. This is presumably to prevent someone from using a grounded appliance on the ungrounded circuit. What's the code requirement, and what about this epoxy business?

A. In my jurisdiction (Oakland, Calif.), two-slot receptacles are readily available and must be maintained on existing ungrounded receptacles unless the outlet is upgraded with a ground per the NEC. The "epoxy business" sounds funky to me. It is not an acceptable practice where I work, and there is certainly no such "fix" in the NEC. This issue most often surfaces when someone wants to install a dedicated computer outlet. In that case, a separate ground may be run, as noted above, and a three-prong outlet installed. — *Redwood Kardon*

To Pigtail or Not To Pigtail?

Q. I have always thought that the best way to wire a receptacle is to use a pigtail lead from the supply wires to the receptacle. My electrician prefers to run the supply wires, and also the wires to the downstream receptacles, to the screw terminals at the back of the receptacle. When receptacles are wired his way, with the feed-through current going through the receptacle, can downstream electrical loads overheat the receptacle?

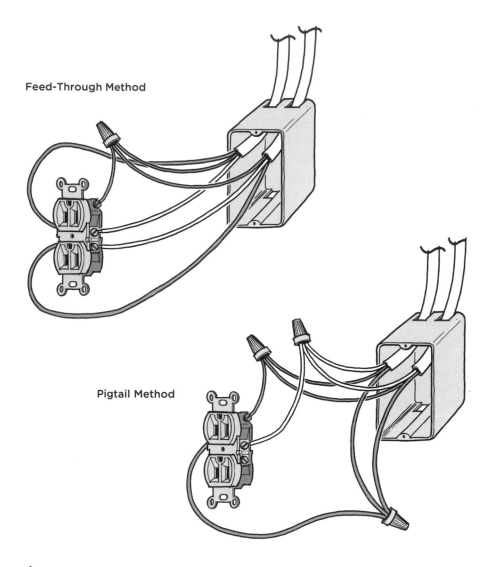

Feed-Through Method

Pigtail Method

A. UL specifically tests receptacles for the ability to safely carry downstream current. This testing is usually conducted at 20 amps, since a 15-amp receptacle can be wired on a 20-amp circuit and thus carry the 20-amp feed-through current. As long as the receptacle is rated and UL-listed for feed-through wiring, as well as properly installed, it should be safe as intended by UL's Standard for Safety (UL 498), which covers receptacles.

The pigtail method transfers the connection point from the receptacle to a wire nut. So instead of the feed-through current going through the receptacle, it goes through the wire nut. Which junction is better? It's hard to say; it depends on the skill of the installer. The methods should work equally well as long as the installation is properly made in accordance with the manufacturer's instructions, and in compliance with all provisions of the National Electrical Code and any UL listing require-ments that may apply. Be aware that in multiwire circuits — circuits shar-ing a neutral — you must pigtail the neutral, as the code does not allow you to use a mechanical device, like a receptacle, to pass neutral current from wire to wire. — *Steve Campolo*

Putting New Devices in Old Walls

Q. What are the choices for putting new switch and outlet boxes in old walls? Any hints for attaching the boxes?

A. Remodel boxes, or remodeler straps, also called "chicken legs" or "turkey legs" (see photograph), are helpful for putting new boxes in old walls. The remodel boxes come in plastic or metal. The plastic ones have ears that lie flat against the box as you place the box into a hole in the wall. Then, by turning screws from the front of the box, the ears turn out from the box and catch the inside of the wall. This places pressure from behind the box, and the flange on the front of the box holds the box firm-ly in place. The metal version works on the same principle, but will not flex or distort as the plastic box can. — *Cliff Thomas*

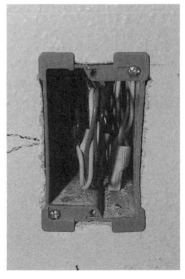

Remodel boxes and remodeler straps (left) are helpful for putting new boxes into old walls. Once in position, the flange on the front of the remodel box holds it firmly in place (right).

Installing Smoke Detectors

Q. There are no code requirements for smoke detectors in my area. What is the best type of smoke detector to buy, and where should they be located?

A. To be sure you are complying with local regulations, contact your local electrical inspector and the local fire department. For new construction, BOCA and the National Electrical Code both require hard-wired, interconnected devices with battery backup.

Interconnected alarms are usually wired in a daisy chain, using 14/3 or 12/3 cable. The third conductor is the communication wire. With the advent of AFCIs, all smoke alarms that are in bedrooms must have AFCI protection, as do all bedroom outlets (lights, fan, etc.)

Some installation guidelines:

- As a minimum, one device is required on every occupiable level of the house.
- At wall-ceiling intersections, there tends to be a dead space where smoke doesn't accumulate, so smoke detectors shouldn't be located in corners. While smoke-detector manufacturers allow the devices to be located as close as 4 inches to the wall-ceiling intersection, some codes require that a ceiling-mounted smoke detector be located at least 12 inches from a wall and that the top of a wall-mounted smoke detector be located 6 to 8 inches from the ceiling.
- Install a separate smoke detector in each bedroom and one in the common hallway near the bedrooms.
- In most cases, a smoke detector should not be located in a kitchen. The best device for use near a kitchen is a smoke detector with a hush mechanism, like the Firex model 4518 or 4618 (Invensys Controls America, 800/951-5526; www.icca.invensys.com) or the First Alert model SA 4121B (800/323-9005; www.firstalert.com). The hush mechanism disables the alarm circuit while keeping the sensing circuit active. If the amount of smoke continues to increase, the device will sound again.
- An attic accessible by stairs (as opposed to a ladder) needs a smoke detector. Although NFPA 70 exempts unfinished attics, an attic with stairs might be used by children to play in.
- A basement is considered an occupiable level and requires a smoke detector. If the stair is open (with no walls), the smoke detector should be installed at the top of the stairs. If the stair is closed (with walls on both sides), the smoke detector should be installed at the basement ceiling near the bottom of the stairs.
- Although not required by most codes, added protection can be achieved by installing a heat detector near the furnace or boiler and

in the garage. These heat detectors should be interconnected with the house smoke detectors.

• Large houses may need additional smoke detectors. Some codes require one smoke detector for every 1,200 square feet.

— *Art Laurenson*

Extending the Depth of an Outlet in Remodeling

Q. When you add a new layer of finish, what is the easiest way to bring the electrical boxes flush to the new wall surface?

A. Arlington Industries, Inc. (800/233-4717, www.aifittings.com) and Lamson & Sessions (800/321-1970, www.lamson-sessions.com) both make an adjustable plastic box extender for just this purpose. Alternatively, a utility (or handy) box extension would provide almost 2 inches of added depth for a single-gang box. For two-gang extensions, an extension ring for a 4-inch square box at 1 1/2 inches deep is also available.

Boxes are allowed to be set back only 1/4 inch if the wall surface is a noncombustible material, such as drywall or tile, but must be flush with the finished wall surface if it is combustible, like wood paneling.

— *Cliff Thomas*

Advantages to 240-Volt Lighting

Q. I am planning to install new lighting in a warehouse. Are there any advantages, either in installation costs or in operating costs, to installing 240-volt lighting fixtures instead of 120-volt?

A. Since you can install twice as many lights on a 20-amp 240-volt circuit as on a 20-amp 120-volt circuit, installation costs for 240-volt lighting should be lower. If your building is large, the savings in labor and materials could be substantial. Another benefit to 240-volt lighting is that the load is automatically balanced; there is no need to balance each 120-volt leg, as there is with 120-volt lighting.

One drawback to using 240-volt equipment is the limited availability of replacement ballasts. If you plan to use HID (high-intensity discharge) lighting, this is not a concern, since most HID lights have multi-tap ballasts. But 240-volt fluorescent ballasts are not widely available and would have to be obtained at an electrical supply house.   — *Sean Kenney*

Finding Buried Outlets

Q. What's the best way to locate an electrical box that was covered over when the drywall was installed?

A. First, try to narrow down the approximate location of the hidden box. The National Electric Code (NEC) requires that outlets be placed no more than 12 feet apart, and that there be an outlet no more than 6 feet from a door opening. Most outlets are located 18 to 24 inches off the floor, and wall switches are generally located about 4 feet above the floor.

Carefully scan the wall surface in these areas for a hump created by the hidden box pushing against the drywall. If you can't see a hump, place a long straightedge on the wall, and move it around until it rocks over the high spot.

I carry two types of electronic locators that help me pinpoint the location of the box. If the hidden box is "hot," I use a "sniffer" to pinpoint the location. If there is no power at the hidden box, then I'll tap into the wires of a downstream box and use a tone generator to locate the hidden box.

After making my best guess at where the center of the box is, I drill a 3/8-inch hole, being careful not to damage the wires by drilling too deep. I shine a flashlight in the hole and verify that I've pinpointed the box. Using short strokes with a keyhole saw, I carefully cut outward from the hole until I reach the walls of the box. Then I cut around the outside of the box, allowing the drywall to draw up tight to the studs. Nearby fasteners will stand proud of the surface and need to be driven home.

Sometimes when the wall has already been finished, I may cut out an undersized box opening and let the bulge remain in the wall. This isn't the prettiest approach, but it eliminates the need to patch the wall, and may work in a utility room or other secondary space. The NEC requires that the face of the box must be flush with the wall surface, so I always install a box extension to satisfy code. — *Sean Kenney*

Concealed Splices in Remodels

Q. What are the rules for dealing with old concealed splices?

A. Any splice that is at all suspect should be checked for mechanical integrity and made code-worthy. All splices are required to be in a box. However, splices cannot be concealed. Once found, they have to be left accessible. — *Cliff Thomas*

Arc-Fault Circuit Interrupters

Q. What is an arc-fault circuit interrupter? Are they expensive? Are they subject to nuisance tripping?

A. According to the National Electrical Code, these little babies are required for all bedroom outlets — not just receptacle outlets. An arc-fault circuit interrupter (AFCI) is like a common 15-amp or 20-amp single-phase circuit breaker with a twist. They work by sensing an arc on the branch circuit, then opening the circuit. Unfortunately, they are subject to nuisance tripping, although newer models work better than the first ones made. — *Rex Cauldwell*

Electrical Fluctuations

Q. I am having an electrical problem on a second-story addition I am currently building. The clients report that the lights dim periodically during the early morning hours, but that it's not related to the refrigerator starting nor to any other piece of electrical equipment in the house. I have tested the voltage and have found no fluctuations greater than two or three volts during the course of a day.

The service is 200 amps, and the entire house has been completely rewired within the past six years. The wiring looks to have been properly installed.

Is there something I have overlooked in trying to determine the cause of these electrical surges? The adjacent houses have not reported any surges.

A. Dimming of lights is not caused by a surge but by a lowering of voltage. When the full voltage and brilliance of the lights return, it just looks like a surge.

I would advise you to hire an electrician skilled in advanced troubleshooting. The problem with situations like these is that you have to be there when the dimming occurs to be able to isolate the cause. I once had to move into a house for a couple of days to troubleshoot an inter-mittent problem.

Most of the time, however, that's not necessary. There's a good chance you can simulate the conditions that are causing the problem. Despite the homeowners' insistence that the dimming is not related to electrical equip-ment (homeowners are often wrong in what they report), the early morn-ing hours are usually a time of heavy electrical usage — hairdryers going, cooking in the kitchen, lots of lights on, the well pump if there is one, and so forth. I've also seen lights dim when a gas dryer was running — the belt was too tight and the motor was having a hard time starting.

First, try to verify whether the utility is responsible for incoming low voltage. To do this, I turn off the breaker for the electric stove — usually the heaviest power draw in the house. I then turn all the cooking ele-ments and the oven on high. With my VO meter probes on the panel's main lugs, I throw on the stove breaker to see if the voltage drops below 240. If it goes down 10% or more, then there's probably a problem at the transformer.

If the voltage remains steady, next test the two phases. Use a portable electric space heater to load first one side, then the other. With the heater running, measure the voltage between each phase and neu-tral. If one phase goes down several volts while the other phase goes up several volts, then most likely the service entrance cable (SEC) neutral is starting to deteriorate and the SE cable will have to be replaced.

Another possibility is that the SEC neutral splice at the panel or the meter has loosened and needs attention. This happens with SE cable because the soft aluminum compresses under the lug and loses contact.

The problem could also be caused by a bad neutral on the utility side as well. This is verified by measuring at the meter base, but you'll have to have the utility there to cut the seal.   — *Rex Cauldwell*

Wiring a Two-Pole GFCI

Q. I have a 240-volt electric jobsite heater, which is required by the NEC to be GFCI protected. Will a 120/240 two-pole GFCI breaker provide protection for a 240-volt load with no neutral? In what situations would a 120/240 two-pole GFCI breaker be recommended?

A. Some equipment, including some spas, is pure 240-volt, with no neutral. Most everything else is both 120- and 240-volt. It doesn't make any difference if the load is pure 240-volt with two hots and no neutral or 120/240-volt with two hots and a neutral — you use the same double-pole GFCI breaker. You even install it the same way. Wire both hot con-ductors to the breaker and the breaker pigtail to the neutral bus. There will be no connection to the breaker neutral, so just ignore it.

The way the breaker works is via "vector addition." It sums the current of the load and uses that as a reference as the current leaves one leg of the breaker. In theory, the current coming back should be the same. If it is not, the breaker opens. The vector addition will not work if the pigtail is not connected.
— *Rex Cauldwell*

Detecting Live Romex

Q. I'm a remodeler and often work around existing wiring. What's the most practical and inexpensive tool for detecting live current in either Romex or wire in a metal conduit? Can you detect current inside the conduit without opening the conduit?

A. In my opinion, the only dependable way to detect live current — the way I would want to bet my life on — is with a multimeter connected to the wires. Check the voltage, conduit to ground, ground to hot, ground to neutral, and neutral to hot. The wands, or "sniffers," I've used in the past — the types you wave over the wires to tell you whether they're hot — have not been dependable. However, the new ones seem to be right on. Regarding wires in conduit, you must remember that the conduit may be grounded, which would give a "dead" reading when the wires inside are hot.

Another problem with low-cost voltage sensor pens is that many don't detect voltages of less than 60 volts. Thus, if there is a low-voltage circuit or a circuit making intermittent contact, the sensor won't pick it up. One unit that I have not tried but that looks promising is made by Extech Instruments (781/890-7440, www.extech.com). Their model DVA30 is claimed to measure from 5 to 250 VAC through conduit or shielded wire.
— *Rex Cauldwell*

Flickering Lights

Q. I'm doing some wood siding repairs on a house, and every time I hit the trigger on my chop saw, the lights in the house flicker. The power company says the problem is with the house, not their lines. But the lights in the house across the street flicker, too. A friend suggested that there could be a bad connection on the neutral. How can I determine what the problem is and then get the power company to fix it if it's their responsibility?

A. This has to be a utility problem: either low voltage or too great a load for the available voltage. There is no other way your chop saw would be able to affect the house across the street. Probably both houses are working on the same transformer, and the transformer is underpowered. If you are on different transformers, then the utility really has a problem. It means that their high-voltage line is overextended — in other words, there are too many homes on that particular tap.

The "bad neutral" theory doesn't add up. A service entrance connection (SEC) neutral problem gives a whole set of different problems. Every house has two energized lines into the house and a neutral that either hot line can use. When an SEC neutral goes bad, one of the hot legs to the house goes high, and the other leg goes low. In the house, that will likely damage anything connected to the leg that goes high.

That's not to say that there isn't also a problem in the house that is making things worse. For example, if a house is underpowered, that would tend to make the lights dim more easily to begin with. The homeowners should have an electrician check out the utility problem and verify it; that way, the utility is hearing from a professional that it is their problem.

Here's how I would troubleshoot:

1. With a high-quality digital volt-ohm meter (VOM), measure the voltage going into the service panel under load and nonload. Load one leg, then the other, then both. The voltage should not change more than a few volts. If it does, the transformer is too small.

2. Make the same measurements with someone across the street operating the chop saw. If the voltage goes down (and we know it will), the problem is outside the house.

3. Make similar measurements at the utility meter. If the voltage drop occurs before going into the house, we know the problem is outside the house.

4. Look up at the transformer and read the kilovolt-amp (KVA) rating, which is often painted on the side. A 200-amp house at 240 volts needs a 45-KVA minimum; with two houses you need twice that. Odds are, they are running both houses off one 45-KVA transformer.

I'm confident that the main problem will turn out to be the power company running too low a voltage on the primary and overextending the tap line. I've seen this more times than I can count. If the utility won't cooperate, the house owner should write the state utility board, with a copy going to the utility. Normally the utility has 30 days to respond to the board. — *Rex Cauldwell*

GFCI Protection for Shower Lighting

Q. In a bathroom remodel, we plan to install a light fixture in a shower stall, with a switch near the shower door. Do the fixture and switch need to be GFCI-protected?

A. The short answer is yes. Although the NEC does not require either the switch or the light fixture to be GFCI-protected, most lighting fixtures designed for use in a shower stall require GFCI protection to meet UL requirements. Providing GFCI protection for a switch within reach of a shower stall is a good safety practice, even when not required by code. It's worth consulting your local inspector before you locate a switch within reach of a shower stall. Even though the NEC allows such an installation, two inspectors have made me move switches so they were out of reach of the shower stall. — *Sean Kenney*

Wiring Bathroom Lights

Q. At several jobs, I have noticed that the electrician wires the bathroom light downstream from the bathroom GFCI receptacle. When the GFCI trips, the light goes out. Does this meet the NEC?

A. Even though it doesn't make sense, I'm afraid that this installation does meet the code. The NEC doesn't require installations to follow common sense or be convenient. Certain dangers are permitted, as long as the danger is non-electrical.

The code doesn't care that if the light goes off, you may be nude, walking in darkness on a wet, slippery floor, trying to find the door handle, or that you may have to walk (presumably, with a towel wrapped around you) through the party in the living room out to the garage to throw the GFCI breaker back on. — *Rex Cauldwell*

Service Panel Surge Protection

Q. Does a surge protector installed in the service panel eliminate the need for individual surge protectors throughout the house?

A. No. If you're serious about protecting electrical equipment, you should use both types of devices. I'd recommend installing a secondary lightning arrester inside the main service panel and providing a transient voltage surge suppression device at the equipment's point of use.

The suppression system in the service panel will handle the surges that enter via the ungrounded conductors feeding the panel (a lightning hit, for example). The point-of-use suppression device will squash spikes that enter through the utility's neutral conductor and also take care of smaller spikes that make it past the service panel protection.

For in-panel protection, I prefer the Siemens all-in-one breaker/surge filter unit (800/964-4114, www.sea.siemens.com) or one of the Tytewadd Power Filters (417/887-3770, www.tytewadd.com).   — *Rex Cauldwell*

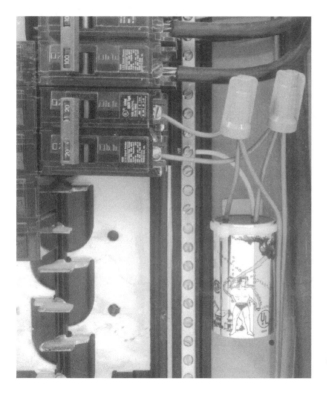

Tytewadd in-panel surge arrester

Thermal & Moisture Control

6

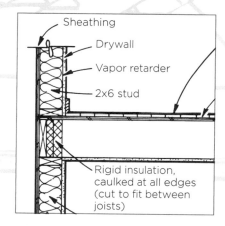

Sheathing

Drywall

Vapor retarder

2x6 stud

Rigid insulation, caulked at all edges (cut to fit between joists)

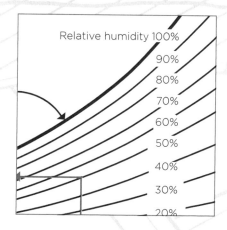

Relative humidity 100%

90%

80%

70%

60%

50%

40%

30%

20%

Venting Bath Fan Into Soffit Not a Good Idea

Q. It has become common practice in new construction around Nashville to vent bathroom fans into a nearby soffit. One problem is that the vent often gets covered with vinyl or metal soffit because the siding crew doesn't want to cut a hole in the finished soffit. Other than running the vent into the attic, which is not a good idea, what can you do? With mold problems becoming more prevalent, we need more options.

A. The International Residential Code (Section R303), referring to mechanical ventilation in the bathroom, requires that "the minimum ventilation rates shall be 50 cfm for intermittent ventilation or 20 cfm for continuous ventilation. Ventilation air from the space shall be exhausted directly to the outside."

This seems pretty clear: venting to the attic is out of the question, as that is not directly to the outside. Venting into the area behind the soffit panel would be the same thing, while venting through a hole in the soffit is only a step away. While using a vent cap mounted in the soffit panel might be considered venting to the "outside," the moist exhausted air will most likely be drawn back into the attic through the soffit vents as soon as it hits the outside (assuming that the soffit and ridge vents are properly sized and installed and have good convective flow). So while this follows the letter of the code, it doesn't follow the spirit. Although it may cost more and add another penetration through the roof, you should take the time to do this right: install a roof vent or, preferably, a gable wall vent, and flash it properly.   — *Harrison McCampbell*

Makeup Air for an Exhaust-Only Ventilation System

Q. I'm building a tight house with above-average attention to air sealing, and I plan to ventilate with a Panasonic exhaust fan running continuously. Will the cracks around windows and doors admit adequate makeup air, or do I need to provide wall vents?

A. Studies have shown that even a tight home usually has enough openings in the building shell to provide makeup air for the base ventilation rate of most homes (45 to 90 cfm). Dedicated passive makeup air inlets have been shown to be ineffective, since the fans used for ventilation typically do not generate the high level of negative pressure (10 to 20 pascals) needed to draw outside air through the inlets.

The main concern for your proposed system is not whether the house has enough cracks for makeup air; it is whether the makeup air will be drawn from the wrong locations. Potentially, an exhaust-only ventilation system can cause backdrafting of open combustion systems (fireplaces, water heaters, furnaces, or boilers), or the entry of soil gases into the home. If you plan to use exhaust-only ventilation, it's important to install a pre-radon mitigation system and to use only sealed-combustion appliances. Before using any open-combustion systems, a worst-case depressurization test of the house should be performed. — *Bill Rock Smith*

Tight Houses and Air Quality

Q. Do tight houses have higher levels of humidity, radon, and indoor air pollution than older, leakier houses?

A. With a strong enough contaminant source, any house can have an indoor air quality problem. If contaminants are properly controlled, the air in a tight house can be as clean as or cleaner than the air in a leaky house.

For example, some of the houses with radon problems I've worked on would need 20 to 30 air changes per hour to get them below the recommended levels for indoor radon. This could be done by adding big enough fans, but the house would cost a fortune to heat and you wouldn't be able to keep candles lit. So the first rule is: "No strong sources of air contaminants in the house."

Some sources, however, like moisture and odors from kitchens, laundries, and bathrooms, are unavoidable. These are actually pretty easy to control by locally exhausting these rooms. Local exhaust is better than a general increase in ventilation because it not only brings in outside air to dilute the contaminant, but it also keeps it from spreading to the rest of the house. ASHRAE recommends 100 cfm of intermittent exhaust in the kitchen and 50 cfm in each bathroom.

Other unavoidable contaminant sources are scattered throughout the building — notably, people. People give off bioeffluent (body odor), as well as carbon dioxide and water vapor from breathing. (With too much CO_2 in the air, you feel drowsy and overheated.) There are also fungi, bacteria, mites, insects, rodents, dogs, and cats — who all give off odors — even in the cleanest of houses. To control these contaminants,

ASHRAE (Standard 62.2) recommends that residences have mechanical ventilation systems capable of providing an air exchange of 7.5 cubic feet per minute (cfm) plus 1 cfm per 100 square feet of occupiable floor area.

The ventilation system should be designed to control the last category of unavoidable contaminants — soil gases, which include water vapor and occasionally radon and methane. Use the air handling equipment to slightly pressurize the basement or use an exhaust fan to depressurize the subslab drainage layer. Similarly, you should design the ventilation system to slightly depressurize the upper parts of the house (in northern climates) to protect the walls and ceilings from moisture condensation. — *Terry Brennan*

Dryer Vent Installation

Q. What's the right way to vent a dryer? The standard plastic flex-hose with the spiral wire always collects condensation and sags. There has to be a better way. Would metal or plastic pipe work?

A. In Massachusetts, where I live, the code won't allow use of the plastic flex pipe if the pipe passes through a wall, and code officials discourage its use in all cases. Also, several dryer manufacturers have disallowed it because it can't stand up to the hot air coming from the dryer.

From a mechanical perspective, it's a poor idea. First, the spiral configuration of the flex pipe creates a lot of friction in the line, slowing the exhaust. Also, the sags and turns typical with flexible pipe further reduce airflow, and allow lint and moisture to accumulate, creating blockages.

I would use a smooth metal duct. You could even insulate it to reduce condensation if you think this will be a problem (if the dryer is in a cold space, for example). You should slightly pitch the line toward the outside if possible to allow any condensation to drain. In any case, you'll have no sags where the moisture can collect. PVC pipe might also work for this application, but here in Massachusetts at least, you would have to provide a manufacturer's temperature rating to the inspector to prove that it could take the heat of the dryer's exhaust. — *Paul Fisette*

Radon Resources

Q. Where can I obtain information about radon gas, its effects, and what products will prevent emissions from coming up through slab foundations?

A. The best place to start is the Environmental Protection Agency's Radon Mitigation Standards Web site at www.epa.gov/radon/pubs/ mitstds.html. There you'll find links to a wide range of useful information, including a searchable index, radon publications, radon hotlines, radon myths and facts, and a good collection of other radon links. The EPA also provides you with directions for finding a qualified radon service professional in your area (www.epa.gov/radon/proficiency.html). Since the EPA shut down its National Radon Proficiency Program (RPP) in 1998, it also provides contacts for every state at www.epa.gov/iaq/where youlive.html.

— Paul Fisette

Insulation

Optimal Insulation Thickness

Q. In a heating climate (central New York), I would like to know how to determine whether adding more insulation results in diminishing returns in terms of energy savings. Is there any reason to include more than R-50 insulation in an attic, assuming the access is good?

A. Every inch of insulation you add to an attic results in less savings than the previous inch you added. That's just the way the physics works out. The good news is that adding another inch of insulation in an attic costs very little, if any, in added labor, so it's mostly added material costs. For a house in central New York (6,700 degree days) with typical efficiency oil heat (75%) and oil at $2.53/gallon, the cost for the heat lost through the insulation in a 1,200-square-foot attic for a typical year is summarized in the table below.

Local codes usually dictate the minimum acceptable level of attic insulation. What kind of savings can a homeowner expect when they upgrade from R-30 to R-40? It would be worth another $41 in the first year. If the homeowner expects a simple 10-year payback, then up to $410 could be spent on the upgrade to R-40 and still be worth it. The added cost for R-38 unfaced batts over R-30 (retail at my lumberyard) is $0.15/sq. ft., or about $180 for the attic, so that looks like a good deal for the owner.

Fuel Cost vs. Attic Insulation R-Value
(6,700-degree-day climate, 1,200-sq.-ft. attic)

Attic insulation R-value	Annual fuel cost for attic heat loss	Annual savings for each R-10 increment of insulation added
10	$475	—
20	$238	$237
30	$160	$78
40	$119	$41
50	$95	$24
60	$79	$16

How about R-50? To get R-50, you'd probably add R-19 to the R-30 batts, at a cost of $0.46/sq. ft. over the R-38 batts alone, or about $552 for the whole attic. The first-year savings is $24, so with a 10-year payback, it's worth $240. That doesn't look like a great deal for the owner. It gets worse for more insulation: the cost to go to R-60 over R-50 yields even less savings.

This is why R-38 batts are so common in attics in heating climates. It's usually the right price point for the owner. If you are blowing in cellulose (my favorite, as it fills all the odd-shaped places well and is a recycled material), the incremental costs are a little different. All this said, I've never heard of an owner who was sorry that he put in too much insulation.

— *Andy Shapiro*

Compressing Fiberglass Batts

Q. What happens to the R-value when you stuff R-19 insulation into a 2x4 stud cavity?

A. What happens is that as you compress fiberglass batt insulation, the R-value per inch goes up, but the overall R-value goes down. For every x% that you compress the material, you lose approximately $1/2$x% of R-value. Refer to the chart below to see how this affects common sizes of insulation. This chart refers specifically to CertainTeed products, but can be used for other similar fiberglass batt insulation products. — *Mike Lacher*

Compressed R-Values for CertainTeed Building Insulation

Nominal Lumber Size	Depth of Cavity	R-15 3½"	R-19 6¼"	R-21 5½"	R-22 6½"	R-25 8"	R-30 8¼"	R-30 10"	R-38 10"	R-38 12"
2x4	3½"	5	14	15	15	—	—	—	—	—
2x6	5½"	—	18	21	20	20	22	21	—	—
2x8	7¼"	—	19	—	22	24	27	26	30	28
2x10	9¼"	—	—	—	25	30	—	29	36	33
2x12	11¼"	—	—	—	—	—	—	30	38	37

Preventing Wind Washing of Attic Insulation

Q. Since ventilated attics can be breezy, doesn't the wind rob heat from the insulated ceiling? Would it help to install housewrap on top of the R-38 fiberglass batt insulation? If the client wanted a storage area, I could install plywood over the housewrap. Similarly, would it help to have housewrap on the back of second-floor kneewalls, which I typically insulate with R-24 batts?

A. Let's answer the easy question first. In my opinion, you should put housewrap on the back side of the second-floor kneewalls in a well-ventilated attic. This would reduce wind washing of the insulation. The installation is roughly in keeping with the standard practice of using housewrap on exterior surfaces.

As for the attic floor, I would not put a housewrap and plywood over the insulation. DuPont, the manufacturer of Tyvek, advises that its Homewrap is not recommended for horizontal applications.

I prefer board sheathing to plywood for the attic floor. Plywood has a relatively low moisture permeance, similar to that of painted drywall, so there could be moisture problems on the underside of the plywood. Boards would be more moisture-permeable, would provide adequate wind protection, and would make a suitable floor for storage.

The drawback to this approach, of course, is that the amount of insulation is limited to the depth of ceiling joists. So, if the storage area is not needed, more insulation in the ceiling is preferable.   — *Don Fugler*

Mice and Insulation

Q. Is there any way to keep mice out of insulation? I am working on a remote cabin in southern Illinois, and the mice seem to come and go at will. I'm getting ready to insulate another room, and I don't want to provide new places for the rodents to nest.

A. This is a million-dollar question without (unfortunately) a million-dollar answer. Effective rodent management is based on exclusion — you have to keep them out. Killing rodents once they have entered the structure doesn't solve the problem, because there are always more rodents to take their places. Also, killing rodents with poison bait inside insulation causes other problems: when they die, they decay and smell really bad! Plus, dead rodents attract insects and other decomposers.

Keeping mice out is a big challenge. They can enter a gap that's only 1/4 inch in diameter. But control is not hopeless. Mice begin to enter homes in search of food and shelter during the early fall (when the weather turns colder). Establishing an active trapping effort around the home during this time will pay big dividends.

Tight construction is also essential. Mice can climb well, so be especially careful to identify and seal all gaps in the construction within 3 feet of grade. Don't ignore gaps in the roof construction, either. Look specifically for plumbing and electrical penetrations, spaces under doors, basement floor drains, and other construction gaps. Fill all gaps with inedible materials like sheet metal, hardware cloth, wire mesh, cement, or plaster. Mice can easily chew through most construction foams and caulks.

It's important to make the area around the house unattractive to nesting mice as well. Remove thick vegetation, piles of junk or clutter, bird-feeder droppings (and other mouse food), and any debris from the area surrounding the home. With a little luck and effort, you should be able to keep your rodent roommates to a bare minimum.   *— Paul Fisette*

Choosing Cellulose Insulation

Q. What are the advantages and disadvantages of dense-pack cellulose versus damp-spray cellulose insulation?

A. The thermal performance and installed costs for dense-pack cellulose and damp-spray cellulose are similar. With both methods, the quality of the job is considered "installer sensitive," since cellulose needs to be installed at the correct density to avoid settling.

The disadvantage to damp-spray cellulose is the addition of moisture to the wall. Because of this added moisture, cellulose manufacturers recommend waiting at least three days before hanging the drywall. Since dense-pack cellulose is installed dry behind a netting or reinforced polyethylene, the drywall can be installed immediately. However, since it takes extra time to install the netting or poly, the total time required for a dense-pack installation can be about the same as a damp-spray job with the drying period.

Dense-pack cellulose is prone to creeping onto the face of the framing or pushing the netting or poly beyond the face of wall, which can make drywall installation difficult. This is rarely a problem with damp-spray, since the excess material is scraped flush with the framing face using a scrubber.
— Bill Rock Smith

Will Cellulose Make the Ceiling Sag?

Q. I am planning to build with 12 inches or more of blown-in cellulose insulation in the ceilings. How much insulation can you put in a ceiling before the unsupported drywall between joists sags? I usually use 1/2-inch drywall with trusses 24 inches on-center.

A. According to USG's *Gypsum Construction Handbook*, cellulose and rock wool insulation weigh about 2.5 to 3 pounds per cubic foot. Fiberglass weighs about 1 pound per cubic foot. With 1/2-inch drywall applied perpendicular to the trusses, the allowable weight load is 1.3 pounds per square foot. This is equal to about 6 inches of cellulose insulation. With 5/8-inch drywall, the allowable load is 2.2 pounds per square foot, or about 10 inches of cellulose.

I have seen many ceilings with 5/8-inch drywall over a 24-inch on-center framing with 12 inches of loose-fill insulation that did not have problems. But if you are planning to use much more than this, the drywall should be hung on strapping spaced 16 inches on-center.

Most sagging in drywall ceilings is caused by improper nailing or moisture. The primary strength of drywall is in the paper facing. As the paper becomes wet, the strength is drastically reduced. Even condensation on the back of the sheet or excessive use of water-based texture finishes can cause problems.   — *Henry Spies*

Drying Time for Spray Cellulose

Q. What is the typical drying time for wet-spray cellulose insulation? What happens if the insulation freezes before it dries? Will it dry after it warms up in the spring?

A. Extensive research has been done on the drying of wet-spray cellulose by the Canada Mortgage and Housing Corporation (CMHC). The CMHC found that if the insulation was applied with the proper amount of water, the moisture content of the wood studs, plate, and sheathing rose considerably during the first 30 days after installation. However, the sheathing and framing dried to near normal levels in two to five months. The presence of a vapor retarder and/or wall ventilation did not seem to affect the drying time or the final moisture levels. A slight amount of fastener corrosion and mold growth did occur, but not enough to be of concern.

In another test, "worst case" conditions were simulated by using wet lumber and very wet cellulose, in a very humid climate (Newfoundland).

In that case, the insulation and wood had not dried after two years. There have been other cases reported of walls not drying out and water dripping out from the baseboards. But in these cases the insulation was too wet when installed, and was trapped between a polyethylene vapor retarder on the inside and foil-faced insulation board on the outside.

Freezing of the wet insulation before it dries will delay the drying process but does not seem to cause any problems. There will be no mold growth or significant corrosion at temperatures below freezing.

— Henry Spies

Rigid Foam Under Scissor Trusses

Q. We're considering installing a layer of 1-inch rigid foam insulation on the underside of a cathedral ceiling framed with scissor trusses. Is the foam a vapor retarder, or will we need to install a poly vapor barrier on the interior side of the foam? Will the foam cause problems when it comes time to finish the drywall?

A. Extruded polystyrene (such as Styrofoam, Foamular, and Amofoam) and aluminum-faced polyisocyanurate rigid insulation (such as Thermax) make an effective vapor retarder if installed on the bottom chord of any style truss. If the joints are sealed with a compatible foil-faced tape, the foam will also perform as an effective air barrier. If the foil tape is applied carefully at the joints, there should be no need for an additional layer of polyethylene.

One inch of foil-faced polyisocyanurate foam has a permeance of 0.05 perm. By contrast, 1 inch of expanded polystyrene or "beadboard" has a permeance of 2 perms; since the International Residential Code defines a vapor retarder as a material with a permeance of 1.0 perm or less, 1 inch of expanded polystyrene would not qualify as a vapor retarder. Those choosing to install interior expanded polystyrene foam in a very cold climate or in a jurisdiction requiring the installation of a vapor retarder should install a 6-mil poly vapor retarder in addition to the beadboard.

I don't see any reason why the foam insulation would cause a problem when installing or finishing the drywall, as long as the drywall is screwed properly to the bottom chord of the truss. Applying 1x3 strapping over the foam will reduce the number of screw misses, however, and creates a dead air space that contributes to a small increase in R-value.

— Henri de Marne

Foam Under Floors

Q. Can I use rigid foam panels sandwiched between subfloor and finish floor to insulate over a crawlspace?

A. This can be done, but I would not recommend it. Movement of anything against polystyrene foam creates a terrible squeaking noise. Remember the last time a foam cooler rubbed against something in the back of your truck?

The best way to insulate a crawlspace is to use foam on the stem walls. However, if you opt to insulate the floor cavity, all piping and ductwork must be insulated also. The best way to do this is to spray the underside of the floor sheathing with closed-cell polyurethane insulation.

— *Henry Spies*

Double-Side Vapor Barrier

Q. As part of the gut remodel of a 1940 house near Houston, Texas, we installed 3/4-inch rigid foam over the exterior wall sheathing, followed by vinyl siding. On the interior, we exposed the 2x4 studs and installed fiberglass batts. Then we installed 1/2-inch foil-faced rigid foam followed by drywall. In that climate, will these "foam sandwich" walls trap moisture?

A. Installing a vapor barrier on both sides of a wall is never a good idea in any climate. In the Houston climate, a vapor barrier should be located on the exterior, so your choice of exterior foam sheathing was a good one. The concern is the foam sheathing you installed on the interior.

The good news is that because of the thermal resistance of the interior foam, the wall cavity will rarely be below the dewpoint temperature of the exterior air. The bad news is that if moisture ever gets into the wall — say, due to a window leak or negative pressure caused by leaky attic ductwork — it won't be able to get out easily.

Should you take the foam sheathing off of the inside? That's a hard question. I say don't. If possible, watch the walls over the next few years; each year, cut open a small hole in several spots and look. If you did a careful job with exterior rain control and window and duct installation, the walls probably won't develop mold. If you get mold, you know what to do. But my advice: don't build a wall this way again.   — *Joe Lstiburek*

Insulating an Exposed-Plank Cathedral Ceiling

Q. What is the best way to insulate a cathedral ceiling with exposed 2x6 T&G boards above the rafters?

A. Assuming the rafters and 2x6 boards are properly sized to handle the roof loads, a simple, cost-effective solution is to use a preassembled panel of expanded polystyrene (EPS) laminated to a single layer of 7/16-inch OSB (see illustration). This is essentially a structural insulated panel (SIP) with one layer of OSB missing. Install a 6-mil polyethylene vapor retarder over the tongue-and-groove roof planks. The EPS panels attach to the roof structure with long spikes. You can then roof over the OSB surface with any conventional sloped roof application. The overall panel thickness varies depending on the desired R-value. A nominal 8-inch panel will provide an R-value of around 30.

An alternative is to have the tongue-and-groove decking laminated to the interior OSB surface of a structural insulated panel. In this case, since the panel itself is designed to carry the roof loads, you can save

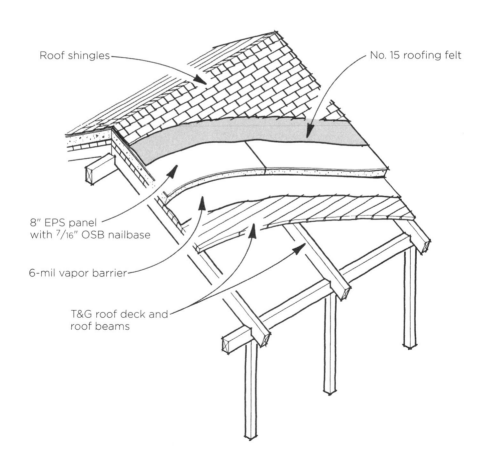

Roof shingles

No. 15 roofing felt

8" EPS panel with 7/16" OSB nailbase

6-mil vapor barrier

T&G roof deck and roof beams

money by using 3/4-inch T&G planks instead of the 1 1/2-inch thickness. These panels are available in 4-foot widths and lengths up to 24 feet (long panels require a crane to set). For a SIP manufacturer in your area, call the Structural Insulated Panel Association at 253/620-7400 (www.sips.org). — *Bill Zoeller*

Patching Icynene Insulation

Q. What is the best material to use to patch Icynene spray foam insulation? Sometimes it's necessary to remove a substantial amount of the stuff after it has been installed.

A. The best way to patch a hole in Icynene insulation will depend on the size of the repair. For small- and medium-sized patches, a can of spray urethane foam (one-component foam from the lumberyard or hardware store) works very well. There is no compatibility problem with using other foam insulation products directly against Icynene.

However, a large patch would require many cans of one-component foam, so it would be better repaired by installing more Icynene or another two-component foam product — applied by a professional applicator. Fiberglass batt insulation would be an option of last resort, since it will not be as resistant to air leakage as Icynene. — *David Ballantyne*

Insulating Old Plaster Walls

Q. What is the best way to insulate and vapor-proof an existing plaster wall without demolishing the surface?

A. It is possible to insulate an existing wall with one of the blown-in insulation materials, such as cellulose, rock wool, or blowing-grade fiberglass. The wall cavities should be blown tightly to prevent settling.

An existing wall cannot be made totally vapor-proof, but in most climates an oil-based paint can prevent any damage from moisture diffusion if indoor humidity levels are kept to reasonably low levels (no condensation on double-glazed windows). Most plaster walls are old enough to have been painted with two or more coats of an oil-based paint already. If not, then two coats should be applied. An enamel is best, even a satin-finish one. The most important step is to seal all openings on the warm side as tightly as possible. This means applying foam sealant around all outlets and other penetrations of the inside surface.

Seal the bottom of the wall where it meets the floor, either with a thin bead of caulk or construction tape. The outside of the wall should not be vapor sealed. The vapor permeability of the outside finish should be at least five times greater than the interior finish. A housewrap or No. 15 felt, which will slow incoming air, will not trap moisture.   — *Henry Spies*

Does Snow Insulate?

Q. Assuming an attic is leaking some heat that reaches the roof sheathing, causing an ice dam, does the snow on the roof tend to act as insulation, trapping the heat near the sheathing? Will the melting of the roof snow be worse when there are 12 inches of fluffy snow on the roof than when there are 4 inches of snow?

A. The simple answer is yes. Snow is an insulator. Its R-value varies, depending on moisture content and density of the snow granules; but on average snow has an R-value of 1 per inch — about the same as wood. Twelve inches of snow have roughly the same insulating value as a 2x4 wall filled with fiberglass insulation.

The worst ice dams occur when there is deep snow followed by a period of very cold weather. To prevent ice dams, the most important and least expensive step is to seal all air leaks connecting the house to the attic. It is also important to reduce conductive heat loss from the house with deep layers of carefully installed insulation.

Once you have minimized heat flow into the attic, install an effective roof venting system. Continuous soffit vents that communicate effectively with continuous ridge vents are a good choice for most houses. I think ridge vents that have an external baffle are best. They provide more reliable suction because the air stream jumps over the baffle regardless of wind direction, creating negative pressure over the vent.   — *Paul Fisette*

Radiant Heat Barriers

Q. I have seen many ads for radiant barriers that are designed to save energy. Is there any evidence that these radiant barriers can reduce home energy costs? If so, in what climates are they most effective? How should they be installed?

A. In cooling climates, radiant barriers can and do save cooling energy. Testing at the Florida Solar Energy Center and other laboratories have consistently shown that a radiant barrier can reduce the amount of heat entering a home through the ceiling by 25% to 40%. The amount of energy saved depends on the level of conventional insulation in the attic. For those with a thick layer of attic insulation, a 40% reduction in the small amount of heat coming through the ceiling is correspondingly small. For those with minimal attic insulation, on the other hand, a 40% reduction in heat flow through the ceiling is a much larger amount.

Two rules of thumb:

- If you have R-30 or better attic insulation, the payback period for the installation of a radiant barrier may be long, although it will save energy.
- In a cooling climate, a house with a radiant barrier and R-19 attic insulation, compared with R-19 with no radiant barrier, should see a reduction in cooling energy requirements of about 10% to 12%. Radiant barriers are not recommended in heating-dominated climates. To my knowledge, there has been no testing of radiant barriers in a heating climate.

The easiest and cheapest way to install a radiant barrier in new construction is to install roof sheathing with a radiant barrier. Several manufacturers now offer OSB or plywood roof sheathing with a laminated radiant barrier. A radiant barrier system can also be installed under the bottom of the top chord of a roof truss or to the bottom edge of rafters. Installing a radiant barrier on an attic floor is not recommended, since such barriers easily get dirty, reducing the performance of the radiant barrier significantly. For more information, contact the Florida Solar Energy Center at 321/638-1015; www.fsec.ucf.edu.   — *David Beal*

What's the Dewpoint?

Q. When people refer to the dewpoint in a wall assembly, are they talking about a location or a temperature? How is the dewpoint calculated?

A. The dewpoint is not a location; it is the temperature at which water will condense out of the air. Since the dewpoint changes with the amount of humidity in the air, as well as the air temperature, the dewpoint for a particular temperature and relative humidity is best looked up in a table or a psychrometric chart (see below).

Water from the air will condense on building components when they are below the dewpoint of the air that's in contact with them. In hot, humid summer weather, water condenses on cold water pipes and drips, as well as on uninsulated basement floors if the space is open to the outside. In an air-conditioned building in a warm, moist climate like the southeastern U.S., the drywall can be below the dewpoint of the outside

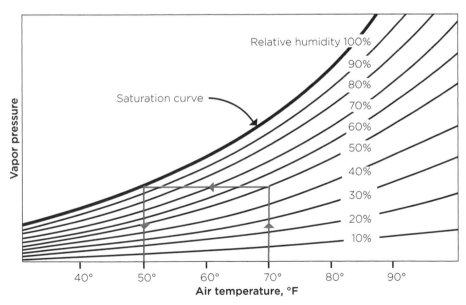

A psychrometric chart provides the dewpoint for any given air temperature and relative humidity. Say you have a relative humidity of 50% at 70°F. On the horizontal scale, locate the air temperature and move up to the curve that represents 50% relative humidity. Then move left to the saturation curve and down to find the dewpoint — 50°F in this case.

air for months on end.

Just because a building component is below the dewpoint doesn't mean there will be a problem. Vinyl window frames and copper tubing aren't bothered by a little moisture. On the other hand, wood window components and drywall can't handle much moisture, especially if the wetting is prolonged and there is no opportunity for the components to dry out.

Determining whether a component in a wall assembly will ever get cold enough to permit condensation — that is, be below the dewpoint — can be complicated. If each element of a wall acted as a solid (which fiberglass doesn't), then the calculation of the temperature at any point in the wall assembly would be fairly easy. Halfway through the insulation value of the wall, the temperature would be halfway between inside and out.

In reality, such static calculations can be misleading, since wall materials can absorb some moisture without being damaged. More accurate calculations, called dynamic calculations, take into account many additional factors but are so complex that they are best performed with computer software. The good news is that this type of dynamic calculation is usually not needed — as long as builders employ good building practices that keep inside air out of walls in cold climates and outside air out of walls in cooling climates, and as long as they also allow building components that occasionally get damp to dry out. One very good source for building details that avoid moisture damage is the Builder's Guide series from Building Science Corp. (978/589-5100; www.buildingscience.com).

— *Andy Shapiro*

Indoor Humidity Sources

Q. A customer with insulated-glass windows has a problem with excessive condensation on the interior of the windows. Can you please list the most common causes of high humidity in a home?

A. Two factors affect the humidity level in a house: how fast water is being introduced and how fast it is leaving. A very tight house doesn't need a lot of moisture input to result in high humidity and condensation on the windows, while the same amount of water introduced into a very leaky house won't raise the humidity much.

You can use a blower door to check the house air leakage rate, though your wet windows may already be telling you that it is relatively tight. You can also check the humidity in the house with a Radio Shack temperature/humidity indicator. The windows should tolerate 40% relative humidity without condensing in cold weather. If they don't, then they are part of the problem — they're not well enough insulated at the edges.

If the homeowner isn't willing to upgrade the windows or add storms, then you'll have to lower humidity levels further.

Showering contributes a lot of moisture to a home. A bath exhaust fan should take care of it, but often the fan is missing, undersized, or little used. Drying clothes indoors also releases a lot of water. (Dryers should always be ducted outside.) Every drop of water that goes to houseplants ends up as moisture in the air. Drying firewood in the basement can add quite a lot of water. Cooking, particularly if the occupants don't use a range hood that is vented to the outside, can generate a lot of moisture.

If the house is new, the construction materials contain literally tons of water that will evaporate over the first winter. Therefore, condensation problems that show up the first winter may not show up again. Poor basement drainage on a wet site can also be a major source of water. (Your nose will tell you if there is water in the basement — you can usually smell the damp or the mold.) Consistently bringing wet or snowy cars into an attached garage that is not adequately sealed from the house can bring in a lot of water as well.

To solve condensation problems, first reduce the sources of moisture and then ventilate to get the humidity down to acceptable levels. I recommend powered ventilation for all houses. An inexpensive ventilation approach is to install a quiet, efficient bathroom exhaust fan, like the Panasonic FV-08VQ. The fan can either run continuously or be wired to a control like the Airetrak, which runs the fan at a constant adjustable speed and has a push button for 20 minutes of high speed.   — *Andy Shapiro*

Mildew in Closets

Q. Why does mildew grow in the closet?

A. To control mildew growth, we must first look at the relationship between air, water vapor, and relative humidity (RH). Water vapor is water in its gas form and is present in air. The warmer the air, the more water vapor it can hold; the colder the air, the less water vapor it can hold. RH is a measure of how much water vapor is in the air compared with the maximum amount of water vapor the air can hold at that temperature. As air cools, without changing its water vapor content, its RH goes up. As the air continues to cool, it reaches the point where the water vapor it contains is all that it can hold — this is 100% RH. Cooling the air any further will result in condensation as some of the water vapor changes to liquid.

Mildew can only grow on surfaces where the RH exceeds 70%. Closet surfaces tend to be colder than adjacent rooms because of poor air circulation from the heated room to the closet, and because relative to their size, they often have more exterior surface area for heat to escape. A corner or cold wall section lacking proper insulation is particularly vulnerable. Since these areas are colder but have just as much moisture in the air as adjacent rooms, they have higher RH and are prone to mold and mildew.

To control the mildew, we have to lower the RH of the closet below 70%. To do this, we either have to raise the closet temperature or lower the amount of water vapor present in the air.

When troubleshooting a closet mold problem, measure temperature and RH in the adjacent room with a sling psychrometer. If the room RH is 30%, then it will be difficult (and hard on the occupants' respiratory systems) to reduce RH much, so look for ways to raise the closet temperature. Either increase heat flow into the closet (in some cases, louvered doors may permit enough heated room air to circulate; in other cases, you may have to put in some heat directly), or cut heat loss from the closet (add insulation, seal air leaks).

If the room RH is 50% or greater, look for ways to reduce the amount of moisture in the air. First, control the moisture at its source. Is it coming from a hot tub, lots of plants, or a gross of gerbils? Is it coming from the six cords of firewood drying in the basement? Next, if the problem occurs in the heating season, increase ventilation levels to replace humid house air with cold outdoor air holding little moisture. As a last resort, use mechanical dehumidification. Closet mildew problems often occur with other moisture problems that the homeowner is not aware of, so a comprehensive "footing to ridge" assessment may be in order.   — *Marc Rosenbaum*

Drying Out a Damp Crawlspace

Q. To dry out damp crawlspaces in Virginia, I have had success closing the crawlspace vents and installing a dehumidifier. Are there any other measures I could be taking? What is an appropriate relative humidity level for a crawlspace?

A. Before beginning work to dry out a crawlspace, there should be a discussion and agreement (ideally, in writing and signed) between the contractor and the homeowner about the potential for house shrinkage as a result of solving the moisture problem. The next step is to confirm that there is not a liquid moisture problem. Is there a live spring, a plumbing leak, a surface drainage problem, or a hydrostatic source for liquid moisture?

Once those potential moisture sources are removed, you will probably want to seal the crawlspace vents. In the Southeast, outside summer air generally holds more moisture in vapor form than the air in the crawlspace does. When this outside air enters a ventilated crawlspace, it can contact a surface cool enough for condensation to form. Any combustion appliances in a sealed crawlspace should be sealed-combustion units, unless ducted exterior combustion air is provided.

Closing the vents and installing a dehumidifier can control the situation if the moisture flow into the crawlspace is less than the capacity of the dehumidifier to remove it. Dehumidifiers installed in a sealed crawlspace do not use a noticeable amount of energy in the summer and generally do not run during the winter. I advise setting a dehumidifier in the 45% to 50% humidity range. In some sealed crawlspace retrofit jobs, dehumidifiers are used only temporarily, until the excess moisture has been removed.

In addition to sealing the vents, you should consider the following steps:

- Install a ground vapor retarder (6-mil polyethylene), sealed at seams and piers with fiberglass scrim and duct-sealing mastic, and held in place with spikes;
- Install a polyethylene vapor retarder on the masonry walls, sealed under the ground poly, leaving a 3-inch termite view strip at the top of the wall; and
- Install weatherstripping to make the access door reasonably airtight.

— Bruce Davis

Exterior Foam and Moisture Problem

Q. If I use foam insulation board on the exterior of a wood-framed building, will it cause condensation within the walls?

A. If your building operates at between 20% and 30% relative humidity (a comfortable level for most people), the walls should be fine with any kind of sheathing.

If indoor humidity rises above 40%, moisture in wall assemblies isn't your only worry. Condensation on windows will probably cause trouble as well, and the mold that starts growing on cold spots can become a health problem for the home's occupants. So your first strategy should always be to reduce indoor humidity.

To be on the safe side, though, it's best to allow for interior humidity approaching 40% or even higher. In that case, exterior insulating foam helps by keeping temperatures within the wall assembly above the dew-point — and the higher the R-value, the better. For moderate indoor humidity, an R-5 to R-7 layer of foam should be okay, but if you're looking at a hot tub room or similar situation, go for R-10 or higher.

To prevent air from forcing its way between sheets of foam, tape the seams with a gap-sealing builder's tape. These tapes hold well to foil-faced foams or extruded polystyrene, but we haven't found a tape that sticks well to expanded polystyrene (beadboard). If you don't tape the seams, cover the building with housewrap. You should also take care to seal electrical outlets and other penetrations. — *Joseph Lstiburek*

Moisture in Post-and-Beam Home

Q. The owners of a post-and-beam we've just completed are having a condensation problem on the inside of the windows. The colder it gets, the more ice develops on the glass, to the point that the windows won't open. We used good-quality windows. The temperature downstairs is 72°F; upstairs it's 62°F. Could the house be too airtight?

A. The problem is that the inside surface of the window is below the dewpoint temperature of the air that comes in contact with it. This means that either you have too much moisture in the house or the window is not insulative enough, so that its inside surface is too cold. You might have both conditions.

The indoor relative humidity (RH) should be around 35%, but it should not go above 50%. Monitor the RH for a week with a device that records high and low readings. You can purchase these at stores like Radio Shack for about $20. If you learn that the humidity in the home is too high, figure out why and lower it. A good place to start is by investigating exhaust fans in bathrooms, kitchens, and other high-moisture areas. Be sure that the fans are working correctly and are being used.

If you find that the RH is okay, I would suspect that the window is not as energy efficient as you think.

Another issue to consider is that post-and-beam houses are built with timbers that often have a high moisture content. Typically, the timbers are installed green, because it's expensive to kiln-dry such large pieces of lumber. This may raise moisture levels inside the house for a year or two.

Because the window units are typically installed within a timber framework involving headers, sills, and side members that are wet, each window has a local moisture source. Get a moisture meter and check the moisture content of the timbers. Eventually, they will stabilize at a moisture content of 8%.

Even if the inside surface of timbers (the side facing the living space) is dry, the surfaces within the wall cavity are most likely wet. If wet timbers are the problem, you can raise the indoor temperature, circulate warm air toward the windows, and use a dehumidifier for the short term.
— *Paul Fisette*

Cold Roof Retrofit

Q. A homeowner in central Pennsylvania has asked me to repair a roof with a history of ice-dam problems. The roof is over a cathedral ceiling that has only 6 inches of fiberglass insulation, with no space for venting above the insulation. Can you provide details for a cold roof retrofit above the existing sheathing?

A. In addition to having ice-dam problems, this roof is poorly insulated and may also have problems with condensation within the unvented cathedral ceiling. Building a cold roof will probably solve the ice-dam problem, but it will neither improve the insulation level nor address the issue of possible condensation.

You should first remove some of the sheathing at the eaves to inspect the insulated space and check for possible water damage caused by the ice dams. Then remove some of the sheathing near the ridge to check for possible damage caused by condensation. Wet insulation should be replaced and wet wood allowed to dry, to prevent further damage and carpenter ant infestation.

If you want to build a cold roof without improving the insulation, first strip the roof down to the existing sheathing. Then install 2x2 sleepers from eaves to ridge, over the existing roof. The sleepers should be nailed or screwed over the existing rafters. They should extend approximately 3 inches beyond the existing fascia to create a new soffit for the installation of standard metal venting strips. Fasten a new fascia to the tails of the 2x2 strapping.

Apply new sheathing over the strapping, followed by No. 15 asphalt felt and roof shingles. An externally baffled ridge vent, such as ShingleVent II, should be installed at the ridge.

A better job would include improving the energy efficiency of the roof, which would also reduce the chance of snow melting. This could be done by adding a layer of 1-inch-thick extruded polystyrene (Styrofoam or Foamular) over the existing sheathing prior to the installation of the sleepers. The rigid foam insulation will also raise the temperature of the lowest level of sheathing, greatly reducing the possibility of condensation problems in the cathedral ceiling. — *Henri de Marne*

Attic Condensation Problem

Q. Last winter, the tenants on the north side of a new duplex we built complained of heavy condensation in the attic. Water dripped on the north side of the attic floor from the underside of the roof sheathing, but nothing like this happened on the south side. The attic is very large, and it is divided by a fire wall. We installed balanced soffit and ridge vents, and all exhaust fans are vented outside. What is causing the condensation?

A. A considerable quantity of moisture is reaching the north attic from below. Once the moisture is in the attic, it condenses on the coldest surface — usually the roof sheathing. Most condensation occurs at night, when the roof is coldest. The problem occurs only on the north side because the sun evaporates the condensation on the south side on a daily basis. The moist air is then vented through the soffit and ridge vents. On the north side, which is not sufficiently warmed during the day, the frost continues to build, until a really warm day comes along and the entire frost coating melts, producing the problem you describe.

During the first winter after construction, a major source of moisture can be the construction materials themselves. With a block foundation, there can be at least a ton of moisture lost from the mortar, wood, paint, and drywall taping compound in the house. In a house with a poured concrete basement, the amount may be several tons. The problem often disappears after the first winter.

Additional moisture may be coming from a damp crawlspace or basement, excessive use of a humidifier, or a gas-burning appliance, such as a water heater or furnace that is not venting properly. A plumbing wall or flue chase can provide a direct path to the attic from the basement if these passages are not sealed at the ceiling level.

You may be able to isolate the problem by "reading" the indoor humidity levels in the house on the windows. If there is condensation on double-glazed windows, the indoor humidity is too high and ventilation fans in the kitchen and bathrooms should be run continuously until the condensation goes away. If condensation occurs on the windows on only one side of the house, the problem may be caused by the use of a humidifier or improper venting of gas-burning appliances. If there is no condensation on the windows, the moisture is probably coming from the basement or crawlspace through the plumbing or flue chases. — *Henry Spies*

Condensation on Ceiling

Q. Often my customers request that we insulate vaulted ceilings when we're reroofing. These roofs usually have 4x6 rafters, 2x6 T&G planks exposed to the inside, and composition shingles on the outside. Our technique has been to strip the roof, lay out 2x4 or 2x6 rafters over the open beam area, install R-30 rigid insulation in the bays, then install plywood and shingle roofing above. Recently, I had a problem with one of these roofs. In an area about 2 feet down from the peak of the living room ceiling, the paint started to peel and mold spots are growing. I checked with a moisture meter and found that water is condensing on the surface of the ceiling. Do I need to somehow vent this assembly?

A. What is probably happening is that interior moisture is being carried by air convection through the joints in the T&G planking, the rigid insulation, and the rafters, and is condensing on the underside of the new

Problem

Moisture condenses on underside of sheathing, and leaks onto exposed ceiling planking

2x4 or 2x6 rafters

Rigid insulation between rafters

Exposed T&G planking and 4x6 rafters

Warm, moist air penetrates through joints in T&G planking and between 2x6 rafters and rigid insulation

Solution

2x3 sleepers on edge create a 2½" air gap

6-mil poly vapor barrier

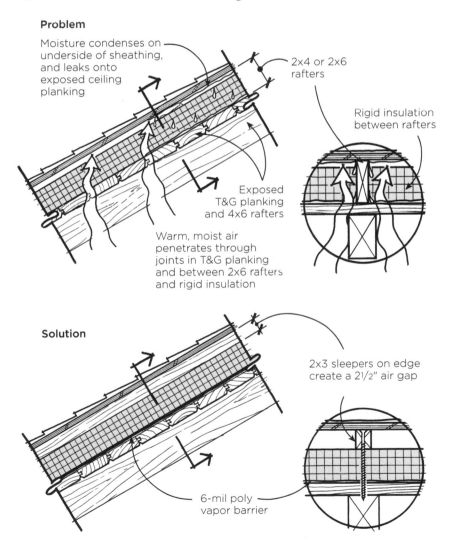

sheathing. From there, it drips down through the joints of the same components back into the living space.

There are some changes you can make to this assembly that should prevent the problem. The first thing you should do is to install a 6-mil plastic vapor retarder over the roof side of the T&G plank ceiling. This will stop the convection of moisture-laden air into the roof cavity. This should be done preferably in one piece. If that's not feasible, start from the bottom and overlap the upper sheets by several inches, and then tape the joints.

The second problem with your assembly is that you are placing the rafters directly over the T&G deck and laying the rigid insulation between them. It would be much more efficient to lay the rigid insulation directly over the decking, thus minimizing the joints between materials and increasing energy efficiency by eliminating the short circuits provided by the rafters. Good-quality extruded polystyrene foam has plenty of compressive strength to handle a roof load.

On future jobs, instead of rafters, use 2x3 sleepers on edge, 24 inches on-center, and fasten them to the deck with long screws. If you mark the location of the beams below, the screws can be driven without concern about coming through the ceiling. With shallower-pitch roofs, a 1-inch penetration into the roof decking should be sufficient as long as you put enough screws in. With steeper pitches, you should provide greater penetration and use strapping across the ridge.

Finally, you should provide continuous ventilation on the underside of the plywood. Extend the tails of the sleepers 3 inches beyond the existing fascia, install a new fascia, and provide soffit ventilation by means of an off-the-shelf ventilation strip. You should also install an externally baffled ridge vent such as ShingleVent II at the ridge. — *Henri de Marne*

Condensation Dripping Under Metal Roof

Q. I designed an open-plan house with what I hoped would be an energy-efficient, maintenance-free roof. The house and its roof have performed well except for occasional annoying ceiling drips, which occur every winter. The roof construction is 2x6 T&G planking, two layers of 3-inch rigid foam, 2x4 horizontal strapping, and a galvanized roof attached with neoprene-gasketed screws. There is a continuous ridge and soffit venting. There is no poly vapor barrier above the plank ceiling because I was assured that the double-layered foam insulation formed its own vapor barrier due to the closed cell structure of the foam. There is no felt above the insulation.

Existing

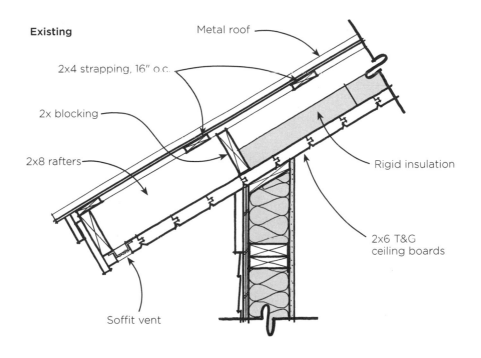

Metal roof

2x4 strapping, 16" o.c.

2x blocking

2x8 rafters

Rigid insulation

2x6 T&G ceiling boards

Soffit vent

Retrofit

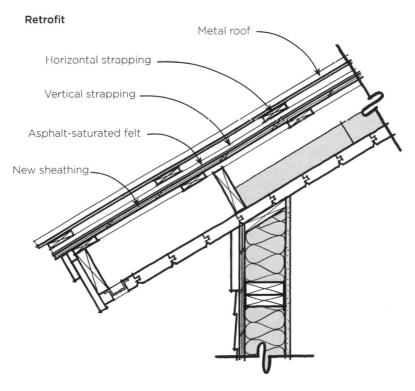

Metal roof

Horizontal strapping

Vertical strapping

Asphalt-saturated felt

New sheathing

Dripping does not occur after heavy rains, but seems to occur most often during thaws after very cold weather, on both north- and south-facing roofs. The water accumulation is a couple of tablespoons full from each of two or three drips. Where is the water coming from — above or below? Is there a fix short of pulling off the roof or putting on a new ceiling?

A. The water is probably condensate forming on the underside of the metal roof on cold nights when night radiation causes the metal to become several degrees colder than the ambient air. The problem may start as frost that melts when the outside temperature rises above freezing or the sun shining on the roof heats the metal. It is also possible that moisture from the interior space is driven through the T&G ceiling boards by convective currents, finds its way through the joints between the rigid insulation panels (which will shrink somewhat as they age), and condenses on the underside of the metal roof. Then, when enough water accumulates, it drips onto the rigid insulation, is blocked by the 2x4 strapping, and finds its way through the joints in the rigid insulation and the ceiling boards. I have seen both of these situations a number of times over the last 20 years.

Had you installed the recommended 6-mil plastic vapor retarder on top of the T&G ceiling, you would have accomplished two things. First, you would have prevented air leakage into the ceiling cavities — air that takes moisture with it. Second, any condensate generated from outside would have been prevented from leaking through the joints between the T&G boards, as long as the plastic were properly overlapped, as is done for roofing felt. However, that water would have been trapped by the 2-by blocking at the eaves and it could eventually have penetrated the walls below or rotted nearby lumber, which would be an invitation for carpenter ants.

The best way to have prevented the problem would have been to install No. 15 or No. 30 felt, which should always be installed under metal roofs, regardless of the type. However, the problem would not have been completely solved unless you had also first installed "vertical" strapping from the eaves to the ridge, so that water would have had channels to drain through.

You mention ridge and soffit vents. How do they work with your strapping system? There is no continuous channel from soffits to ridge (see illustration, previous page). The best solution is to remove the metal panels and nail new sheathing over the existing horizontal strapping. Then apply No. 15 or No. 30 asphalt-saturated felt and strap vertically, then horizontally, to ensure a vent space. Though this may seem like a lot of extra work, screwing the metal directly to the sheathing through the felt would not allow the inevitable moisture that will form on the underside of the metal to evaporate. Instead, the condensate would wet the felt, and could pass through the sheathing around the roof's fasteners where it could eventually cause problems. — *Henri de Marne*

What Is Dry Rot?

Q. I'm confused by the term "dry rot." It seems contradictory since rot occurs when wood is wet. Or is there a kind of rot that happens to wood that is too dry?

A. You are right; rot does indeed require water. Wood rots, or decays, when fungus organisms eat it. Three conditions must be present for the fungus to thrive: temperatures between about 40°F and 100°F, food (which is the wood itself), and a wood moisture content above 20% to 25%.

　　The presence of the fungus is a given because fungus spores are everywhere, carried about by the wind. Two common types of wood-eating fungus are "cubic brown rot" and "white rot." Cubic brown rot eats the cellulose component of wood, leaving the darker brown lignin component, which shrinks into characteristic blocky formations. White rot eats the lignin, leaving the light gray cellulose and covering the wood surface with a white mat of fungal fibers. Both types of rot are found throughout most of North America.

　　Rotten wood is often found in a dry condition and so is called "dry rot." But the wood had to have once been wet for the decay to occur. Another possible confusion arises with "water syphoning fungus." This fungus spreads by forming tubes through which it carries water to wood that is too dry. Syphoning fungus is common in the tropics and sometimes appears in the southern states along the Gulf Coast.　　*— John Leeke*

Identifying Poria Fungus

Q. While renovating a home in New Jersey, I found a fungus-like growth a few inches long growing between the tile floor and the base molding in the bathroom. The house is on a concrete slab. Is there anywhere I can have this growth tested to see if it is poria?

A. Yes, there are laboratories that can identify fungi based on culturing samples taken from decayed wood on site, but the process can be time-consuming and the service may not be free. One laboratory that provides this service is the Biodeterioration Program, Department of Wood Science & Engineering, at Oregon State University in Corvallis (541/737-4222). Should you decide to pursue this, call for specifics on how to collect and prepare the sample for shipment.

　　Because the laboratory procedure can take a month or longer to complete, it may be safer for you to diagnose it in the field. *Meruliporia incrassata* (the scientific name for poria) can be identified by the pres-

The rhizomorph of the porio fungus is a rootlike tube.

One identifier of Meruliporia incrassata *is its fruiting body.*

ence of the water-conducting rootlike tube called a rhizomorph and by the appearance of the fruiting body and the dark colored spores it produces. The rhizomorph looks like a barkless root and smells like a mushroom if broken open.

Since the required repair scenario for homes damaged by a water-conducting decay fungus is different from that for non-water-conducting decay fungi, it is critical to determine whether or not the damage has been caused by poria. For non-water-conducting decay fungi, you have to find the source of liquid water that is making the wood wet enough for decay to occur, and stop it. With poria, the rhizomorphs supply the water, and therefore they all need to be found and severed. Severing the rhizomorphs will eliminate the water source and kill the fungus.

Finding and cutting the rhizomorphs can be difficult, particularly with slab-on-grade construction. The rhizomorphs can move through some very tight openings, such as plumbing penetrations and cold joints between two concrete pours. To prevent poria from infecting the house again, you need to change the construction detail that allowed the rhizomorph to get to the house without drying out (for example, by providing an adequate air gap between wood and soil). Because of the importance of finding the rhizomorphs, you should consider enlisting the help of a pest control operator with poria experience.

The University of California Center for Forestry has prepared a poria package with a summary of many of the published documents on *Meruliporia incrassata.* To request this package, call 510/665-3580 and leave your name and address after the recorded message.   — *Stephen Quarles*

Ceiling Vapor Barrier — Yes or No?

Q. Should you put a vapor barrier in an insulated ceiling or not? I build in a cold climate, where many experienced builders swear that you shouldn't put a ceiling vapor barrier in. The reasons go something like, "because you have to let the moisture escape," or "because the house has to breathe out the top." What do the experts say?

A. To heck with the experts — here's my answer. Plastic vapor barriers should only be installed in vented attics in climates with more than 8,000 heating degree-days. You can forego the plastic and use a vapor retarder (kraft-faced insulation or latex ceiling paint) in all other climates except

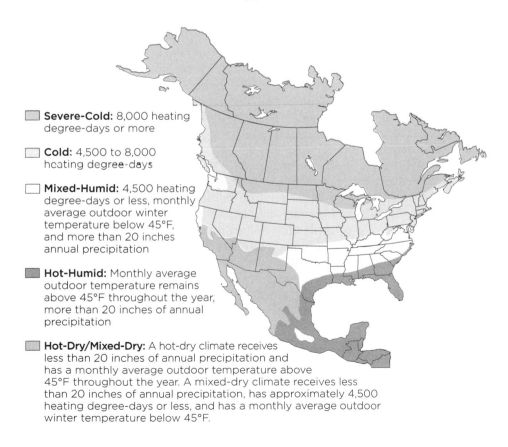

Severe-Cold: 8,000 heating degree-days or more

Cold: 4,500 to 8,000 heating degree-days

Mixed-Humid: 4,500 heating degree-days or less, monthly average outdoor winter temperature below 45°F, and more than 20 inches annual precipitation

Hot-Humid: Monthly average outdoor temperature remains above 45°F throughout the year, more than 20 inches of annual precipitation

Hot-Dry/Mixed-Dry: A hot-dry climate receives less than 20 inches of annual precipitation and has a monthly average outdoor temperature above 45°F throughout the year. A mixed-dry climate receives less than 20 inches of annual precipitation, has approximately 4,500 heating degree-days or less, and has a monthly average outdoor winter temperature below 45°F.

hot-humid or hot-dry climates. In hot-humid climates, attics should not be vented and vapor retarders should not be installed on the interior of assemblies.

In hot-dry climates a vapor retarder should also not be installed, but attics can be vented. All attics — vented or unvented — should have an air barrier (a properly detailed airtight drywall ceiling, for example) regardless of climate.

Omitting a ceiling vapor barrier by arguing that "you have to let the moisture escape" or "because the house has to breathe out the top" is actually correct, in a way. It's also incorrect, in a way. Now, I'm a real believer in controlled mechanical ventilation to limit interior moisture levels in cold and mixed climates, as well as to limit other interior contaminants in all climates. In other words, all houses require controlled mechanical ventilation in order to "breathe." It is also my view that this necessary air change should not happen because of a leaky attic ceiling, attic vents, or even leaky walls. Hence the requirement for an air barrier and controlled mechanical ventilation in all houses regardless of climate.

Having said that, I do not have a problem with relieving some of the moisture load in the house via diffusion. This can be achieved through a roof assembly designed to handle it, such as a vented attic in a moderately cold or mixed climate.

It's important to understand that this is a climate-specific recommendation. In a well-insulated attic in a very cold climate (more than 8,000 heating degree-days), there is not enough heat loss into an attic from the house to allow for much moisture removal through ventilation. That's because attic ventilation requires heat loss to remove moisture from attics. Cold air can't hold much moisture. So ventilating a heavily insulated attic with outside air when it is really cold does not remove moisture. We do not want any moisture to get into an attic in a severely cold climate for this reason.

As you move south into regions where it is not so miserably cold, this changes: hence, the recommendation for a vapor barrier in a severely cold climate but only a vapor retarder in most other locations.

In the old days in severely cold climates, attics were poorly insulated so it was okay to omit a plastic ceiling vapor barrier. The heat loss from the house warmed the attic sufficiently to allow attic ventilation to remove moisture from the attic. Cold outside air was brought into the attic and warmed up by the escaping heat loss, giving this air the capacity to pick up moisture from the attic and carry it to the exterior. This worked well until we added large quantities of attic insulation. With the added insulation, the attic stayed cold and so did the ventilating air from outside, which was now unable to effectively remove attic moisture. Hence the need to reduce moisture flow into the attic and the need for a vapor barrier.

There's one other important qualification: vapor moves either by diffusion through materials or by air leakage through gaps and holes in building assemblies. Between the two, air leakage moves far more moisture than vapor diffusion. A vapor barrier in an attic assembly in a severely cold climate with the absence of an air barrier will likely be ineffective. On the other hand, an air barrier (a properly detailed airtight drywall ceiling, for example) in the absence of a vapor barrier can be effective, since it stops the flow of vapor-laden air. You can't just install plastic in a ceiling and assume it is also an air barrier. For plastic to be an air barrier, it needs to be continuous, meaning all joints and penetrations must be taped or caulked.

— *Joe Lstiburek*

Vapor Barriers in Mild Climates

Q. In a mild climate, should the vapor barrier be on the interior or the exterior of the walls?

A. Codes in many states with mild climates require a vapor barrier on the side which is warm in winter — the inside. However, it is important to remember that walls built in this climate must have the potential to dry.

Mild (or "mixed") climate zones are those areas requiring both heating and cooling for several months of the year. During hot, humid, summer months, the predominant vapor drive will be from the outside in during the sunny daytime. At night, the wall should dry to the outside, if given the opportunity.

An exterior sheathing of plywood or OSB will act as a mild vapor retarder, but will still allow the wall to dry to the outside at night.

To fully understand why this is important, you must understand what a vapor barrier does. First, let's be technically accurate and call this a vapor diffusion retarder! A vapor shield, such as polyethylene sheeting, can only slow the flow of vapor diffusion — it can't stop it.

Vapor retarders are intended to limit the amount of moisture — in vapor form — that passes through the building. The rate of vapor diffusion is determined by the permeability of the building material and the driving force, which is vapor pressure. The higher the vapor pressure or the lower the permeability, the greater the vapor diffusion will be.

Vapor diffusion is also a function of surface area. A vapor retarder that has 30% holes would be only 70% effective. However, this may not be as critical as you think.

Diffusion is only one way moisture migrates through a wall, and the least important. The others, in order of importance, are bulk moisture (rain and snow), capillary action (wicking), and air-transported moisture (leakage of humid air). Thus, having a few small tears in the vapor retarder of a wall is less important than sealing all the wall penetrations, such as electrical outlets, light switches, and gaps between the interior wallboard and the framing.

The bottom line is this: given the opportunity, most conventional wall systems are somewhat forgiving. Control the moisture in the order of importance, and you should have no problem.   — *Frank Vigil*

Housewrap in Hot, Humid Climates

Q. I'm in discussions with an architect regarding the pros and cons of using Grace Ice and Water Shield or a similar membrane material as a housewrap. These materials are more expensive than Tyvek or No. 15 felt but can't be beat when it comes to the wind-driven rain we get here in the Florida Panhandle. Will this work as a housewrap in a cooling climate like ours?

A. Climate has an impact on the specification of a weather barrier system. I would not recommend using a membrane like Grace's rubberized product over wall sheathing in a heating climate. It is impermeable, and it forms a powerful cold-side vapor barrier. If water or vapor leaks into a wall cavity that has a tight "rubberized" covering on the exterior surface and an impermeable vapor retarder installed on the interior surface of the wall, it will be trapped there.

But in your hot, humid climate, a low-perm membrane applied facing the warm exterior could work. It is important to build the wall so that it's able to dry toward the inside when conditions dictate — in other words, don't apply a poly vapor barrier or vinyl wallpaper to the interior.

However, given the expense of rubberized membranes, I think I would choose No. 30 felt. It's stiff and a bit difficult to work with, but it's more forgiving if you get water on the wrong side of the membrane.

If you are building in a very exposed location where "sideways" rain is common, I would consider a rain-screen design, which balances air pressure and creates a drainage plane. This option is not cheap, but I think it's more effective for severe exposures.   — *Paul Fisette*

Vapor Barriers and Insulation at Band Joists

Q. Is it necessary to make the vapor barrier and insulation continuous at the band joist?

A. This detail has confounded builders since the dawn of energy-efficient construction.

Like other elements of the building envelope, the band joist area needs an air barrier and a layer of thermal insulation. In fact, this area

Non-Insulative Sheathing

With standard platform framing, cut rigid insulation to fit between the joists and caulk each section in place.

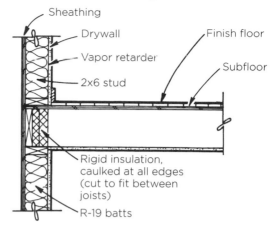

Sheathing
Drywall
Finish floor
Vapor retarder
Subfloor
2x6 stud
Rigid insulation, caulked at all edges (cut to fit between joists)
R-19 batts

Exterior Foam Sheathing

With exterior insulating sheathing, offset the band joist 2 inches and install long strips of foam on the exterior (this keeps the band joist warm enough so that moisture will not condense on it). Some framers, however, find this awkward.

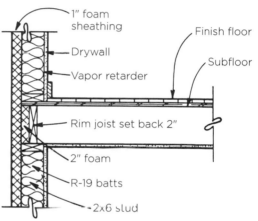

1" foam sheathing
Finish floor
Drywall
Subfloor
Vapor retarder
Rim joist set back 2"
2" foam
R-19 batts
2x6 stud

Interior Foam Sheathing

With interior foam sheathing, the foam board can be easily notched and slid up between the joists.

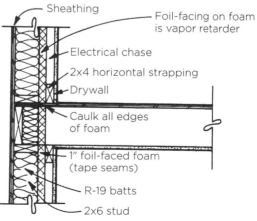

Sheathing
Foil-facing on foam is vapor retarder
Electrical chase
2x4 horizontal strapping
Drywall
Caulk all edges of foam
1" foil-faced foam (tape seams)
R-19 batts
2x6 stud

may be particularly vulnerable to heat loss due to ductwork in the joist system and the fact that the warmest air is likely to lie on the ceiling, which is often penetrated by electrical boxes and recessed lights.

To control water vapor, a continuous air barrier on the warm side of a wall is desirable, since most moisture exits a building with leaking air. The barrier keeps moisture-laden air from getting into wall and ceiling cavities where the moisture can condense on any cold surface. The barrier can be poly, drywall, rigid foam, or any material that will stop airflow.

The best (although not the cheapest) way to insulate a band joist is with medium-density spray polyurethane foam, installed from the interior. Creating a perfect barrier at the band joist — particularly where it runs perpendicular to the floor joists — is probably impossible. As with the rest of the building shell, it is less critical if excess household moisture is removed by mechanical ventilation. Details that I've used successfully are illustrated on the previous page.   — *Chuck Silver*

Foam Sheathing and Moisture Problems

Q. Can exterior foam sheathing cause moisture problems?

A. Exterior rigid foam sheathing is basically a high R-value vapor retarder on the outside of a wood-framed wall. In nearly 20 years' experience, there have been no documented cases in which exterior foam sheathing was solely responsible for damage due to moisture accumulation within walls — whether in warm or cold climates. In fact, a controlled experiment at the U.S. Forest Products Laboratory, in Madison, Wis., showed that walls insulated with foam sheathing had less moisture accumulation than similar walls sheathed with plywood. Evidently, the high R-value foam reduces the condensation potential by keeping the stud cavity warm. With R-11 batts in the wall, use a minimum of 1 inch of foam in cold climates; 2 inches is safer. Combined with proper air and vapor barriers, foam sheathing actually appears to be effective in controlling condensation.

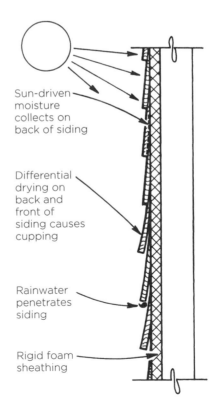

Sun-driven moisture collects on back of siding

Differential drying on back and front of siding causes cupping

Rainwater penetrates siding

Rigid foam sheathing

Wood siding over foam is another issue. There have been plenty of reported failures of wood siding installed over foam without an intervening air space, but not much agreement over whether the cause was the nonabsorbent foam or poor quality siding. The answer may be both.

With vertical-grained, all-heartwood siding that is properly preserved and nailed, the sheathing type probably makes no difference. The siding should have little tendency to deform and the nails should resist minor warping force.

But with poor quality, flat-grained siding that is poorly preserved and installed, the situation is different. The siding may have a greater tendency to warp, particularly if up against nonabsorbent foam sheathing.

The "official" solution to this issue is to select "good quality" siding (if you can get it) and install it properly. But the unofficial and probably safest recommendation is to install 1x3-inch vertical nailers over the foam at each stud. Not only does this create a 3/4-inch cavity to allow backside drying of the siding, but it also provides a much better nail base.

— *Joel (Ned) Nisson*

Vapor-Retarder Paints

Q. Do "vapor-retarder paints" really work? Can I rely on them as the sole vapor barrier in an old house? If so, can you recommend some brand names?

A. Vapor-retarder paints definitely work when applied on the warm side of insulated walls and are especially well-suited for existing homes without an installed vapor barrier. The accompanying product list includes several available brand-name paints.

Vapor-retarder paints typically have a perm rating of 0.8 to 0.45, but the actual perm in the field depends on the number of coats and the degree of coverage. A material's perm rating indicates the ability to diffuse moisture through the material. The lower the perm rating, the better a material's resistance to moisture diffusion. Any material with a perm rating of 1.0 or less is considered a vapor retarder.

While a vapor-retarder paint doesn't have a perm rating as low as 6-mil poly (at about .06), it still qualifies as a vapor retarder, and if applied properly, the paint will slow the transmission of vapor through walls and ceilings sufficiently for most situations. Proper application means full coverage of at least one uniform coat (two coats would be better).

Vapor-retarder paints do have some limitations, however. In extremely cold climates, such as you find near Fairbanks, Alaska, or on the Canadian plains, you'll need to use a better vapor retarder than a coat of paint, especially if the building's interior is above 50% relative humidity. Also, don't forget that any vapor retarder only slows the diffusion of vapor — the transfer of moisture through tiny pores in the wall and ceiling materials. You still need a good air barrier to prevent air from leaking through cracks in the building, carrying moisture into wall and ceiling cavities.

— Joel (Ned) Nisson

Low-Permeability Paints

Aspen Paints
206/682-4603
401 Vapor Guard latex wall primer

Benjamin Moore
201/573-9600
www.benjaminmoore.com
260-00 Super Spec Vapor Barrier latex wall primer

Miller Paint Co.
503/255-0190
www.millerpaint.com
1545 Vapor Lok latex wall primer

Palmer Industries, Inc.
301/898-7848
www.palmerindustriesinc.com
86001-Seal Vapor Barrier non-toxic modified latex primer

Rodda Paint
503/245-0788
www.roddapaint.com
507901 Vapor Block latex wall primer

Sherwin-Williams
www.sherwin-williams.com
See Yellow Pages, "Paint-Retail," for local distributors
Vapor Barrier 154-6407 latex wall primer

Note: This list includes common low-permeability latex wall primers suitable for interior use. In addition to these products, just about any alkyd paint or a pigmented shellac sealer can be used to create an effective vapor retarder.

HVAC

7

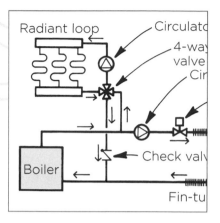

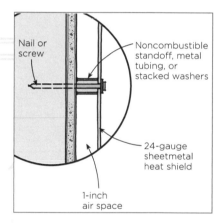

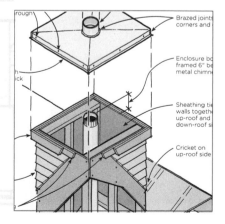

Comparing Propane and Oil

Q. Which home heating fuel is usually cheaper, oil or propane?

If oil costs this much per gallon	Then the price per gallon of propane would need to be lower than this to be a better value than oil
$2.10	$1.39
$2.15	$1.42
$2.20	$1.45
$2.25	$1.49
$2.30	$1.52
$2.35	$1.55
$2.40	$1.58
$2.45	$1.62
$2.50	$1.65
$2.55	$1.68
$2.60	$1.72
$2.65	$1.75
$2.70	$1.78
$2.75	$1.82
$2.80	$1.85
$2.85	$1.88
$2.90	$1.91
$2.95	$1.95
$3.00	$1.98

A. The short answer is that oil is almost always cheaper. But simply comparing the per-gallon prices of the two fuels does not provide a fair comparison, since a gallon of propane has a much lower thermal value (91,600 Btu per gallon) than oil (139,000 Btu per gallon). Remember that a fair comparison must take into account the heating efficiency (AFUE) of the furnaces or boilers being compared; these may be different, depending on the equipment you are considering. Finally, prices vary regionally. The following table shows how cheap propane must be in order to match the price of oil, assuming heating equipment of the same efficiency.

— *Martin Holladay*

Fuel Cost Comparison

Q. A customer asked me how to compare the cost of different heating fuels (natural gas, propane, oil, and electricity). Can you provide a chart with this information?

A. The tables below are based on seasonal delivered efficiency. Because it accounts for equipment, pipe, duct, and other losses, seasonal heating efficiency is lower (by about 5% to 15%) than the listed efficiency of the equipment (the AFUE or annual fuel utilization efficiency) or the steady-state tested efficiency. *— Andy Shapiro*

Oil

Cost per gallon	50% efficiency	70% efficiency	85% efficiency
$2.40	$35.30	$25.20	$20.76
$2.60	$38.25	$27.30	$22.49
$2.80	$41.19	$29.40	$24.22
$3.00	$44.13	$31.50	$25.95

Propane

Cost per gallon	50% efficiency	75% efficiency	90% efficiency
$2.00	$42.78	$28.52	$23.77
$2.50	$53.48	$35.65	$29.71
$3.00	$64.18	$42.78	$35.66
$3.50	$74.87	$49.91	$41.60

Natural Gas

Cost per ccf or therm	50% efficiency	75% efficiency	90% efficiency
$1.20	$24.00	$16.00	$13.33
$1.40	$28.00	$18.66	$15.56
$1.60	$32.00	$21.33	$17.77

Electricity

Cost per kWh	90% eff.	100% eff.	200% eff.	300% eff.
$0.08	$26.04	$23.44	$11.72	$ 7.81
$0.11	$35.81	$32.23	$16.11	$10.74
$0.15	$48.83	$43.95	$21.97	$14.65
$0.18	$58.60	$52.74	$26.36	$17.58

Electric Radiant Heat for a Small Space

Q. I'm considering using electric radiant heating panels in the ceiling of a small superinsulated apartment, but I hear a lot of bias against electric heat. Can it be cost-effective in this case?

A. Electric radiant heat for a superinsulated small space makes a lot of sense. It gives you a low installation cost and a moderate energy cost; it's highly zonable; and there's no combustion. I assume that the hot water would also be provided by electric resistance. An alternative might be a combination wall-hung, sealed-combustion, gas-fired water heater coupled to a wall-mounted kickspace heater or two for heat, but that would cost more to install, take up more space, be noisier, and require the location of a vent and a fuel tank. — *Marc Rosenbaum*

Oversized Heating Systems

Q. Does an oversized heating system waste energy?

A. Yes, an oversized heating system is wasteful for three reasons:
1) It reduces the annual efficiency of combustion heating systems;
2) It increases the potential for flue condensation in mid-efficiency systems; and
3) It increases the cost of heating systems.

Now for some explanation. First, an oversized combustion heating system will not fire as much as a system that is sized properly. Instead, it will "stand by" more and lose more heat up the flue. The more oversized it is, the greater the stand-by time, and the higher the fuel bill.

Off-cycle losses are greater from a boiler (which heats water) than from a furnace (which heats air) because water stores over 3,000 times more heat energy than air for a given volume. This means that boilers are penalized more than furnaces when it comes to oversizing. Also, note that the higher the efficiency of a combustion heating system, the smaller the penalty for oversizing. (The efficiency of electric heat is not affected by oversizing.)

Second, if a combustion heating system runs less because of oversizing, the flue may not stay warm enough to evaporate flue-gas condensation. This could lead to corrosion of the flue, which is a maintenance problem and could result in flue gases spilling into the house.

Third, the larger the heating system, the more costly it will be to install (for any type of heating system).

Studies indicate that the average combustion heating system in the

U.S. is oversized by 2.3 times. Using a conservative estimate of a 5% penalty for oversizing in gas appliances (up to 10% for oil-fired equipment), the savings from accurate sizing is substantial.

Heating systems should be sized for new and existing homes by using design heat load calculations — not rules of thumb or intuitive guesses. — *Richard Karg*

Gas Appliances in Basements

Q. I am remodeling a single-family house with a full basement. The furnace and water heater are located in the basement. The local fire marshal said that it's okay to locate propane appliances such as furnaces and water heaters in the basement. He cites the International Residential Code (Chapter 24) and the National Fire Prevention Association code.

The local building inspector, however, cites the Uniform Plumbing Code, which he says disallows liquid-fired appliances below grade, as in a basement. UPC 1213.6 is specific only to water heaters. Who is right?

A. Sounds like the fire marshal and the code official need to chat. In most municipalities, the building official has the final say when it comes to residential property modifications. Additionally, the codes the fire marshal cites may not be adopted in your town.

The 2000 Uniform Plumbing Code, section 1213.6, does indeed prohibit water heaters from being installed in a basement or pit where "heavier than air gas might collect and form a flammable mixture." The 2000 Uniform Mechanical Code, section 304.6, states about the same for all liquid-fuel gas appliances. I have known some jurisdictions to allow the installation in basements as long as there are provisions to allow unburned gas to drain away (like a pipe) and/or an automatic sensor and gas shutoff. You might ask the building official whether he would approve an alternative method like that. — *Mike Casey*

Indoor Design Temperature

Q. Some HVAC contractors in this area of Virginia size cooling systems using an indoor design temperature of 78°F. I'm under the impression that a system should be designed for an indoor temperature of 75°F. Is there a standard interior design temperature for cooling systems?

A. Both the ASHRAE residential load calculation method described in the 1997 ASHRAE Fundamentals Handbook and the ACCA Manual J are based on an indoor temperature of 75°F. I am not aware of a system-sizing method that uses a 78°F indoor design temperature.   — *Martin Weiland*

Keeping Attics Cool

Q. A two-story house I'm working on has a zoned air-conditioning system with ductwork running through the hot attic, and the A/C can't keep the upstairs zone cool on hot, sunny days. What's the best way to cool that attic space down — add gable vents to the existing soffit and ridge vents setup, insulate between the rafters, or apply a radiant barrier foil under the rafters?

A. Regardless of which strategy you adopt, your first step should be to inspect the attic ductwork. All duct seams should be well sealed with mastic. If the existing ducts are poorly insulated (R-4 or less), upgrade to ductwork with R-6 or R-8 insulation.

We've done instrumented studies of all those options in Florida test houses and test cells, as well as a fourth option you didn't mention: a reflective roof (white tile or white metal, for instance). Here's how the choices compare:

Insulated roof deck. Insulation between the rafters (roof-deck insulation), often advocated without attic ventilation, is an effective way to control heat transfer at the roof itself and to get the duct system into a more hospitable environment. We've measured cooling energy savings of about 9% overall when compared with a standard ventilated attic with an identical roofing system.

But since all the heat transfer is halted at the roofline, it is important to try to control the roof surface temperature to substantially reduce cooling loads (typically by choosing lighter tile or more reflective metal or white shingle roofing systems). With dark shingles, our measurements indicate that insulating the deck will elevate peak shingle temperatures by about 7°F and increase peak decking temperatures by about 20°F.

Although that temperature rise is less than the difference one will see geographically from, say, Detroit to Las Vegas, shingle manufacturers may not warranty their products when installed that way. And remember, those who have studied insulated unvented roof-deck systems do not recommend their use in climate zones where the average monthly outdoor air temperature is lower than 45°F (because of the potential for condensation within the roofing system).

Ventilation. Adding additional attic ventilation will reduce the upstairs cooling load a bit but is likely the least effective option, since the incoming ventilation air is hottest just when you need cooling of the attic. Our experience suggests that adding attic ventilation may produce a 5% reduction to the cooling load. Be aware, too, that attic ventilation can be a double-edged sword. In humid locations during nighttime, it can make the attic more humid. Finally, using powered attic vent fans can further decrease attic temperatures, but the fans typically consume more electricity (300 to 500 watts) than they save on reduced air-conditioning power.

Radiant barrier insulation. A foil radiant barrier material stapled under the rafters can be effective in a case such as yours. The radiant foil is not too expensive (maybe $500 for materials, depending on the roof), and we have measured a 26% reduction in heat flux with the application of radiant foil below a dark-shingled roof. You can increase ventilation at the same time: the heat flux reduction was 36% when we increased the attic ventilation ratio from 1/300 to 1/150 in addition to the foil retrofit. The greatest benefit from a radiant barrier comes when you most need it, with a 16% reduction in A/C demand during the hottest hours from the radiant barrier alone. Again, watch out for excessive ventilation in humid climates, where it can lead to condensation problems on duct systems.

Reflective roof. In new construction, or if you are contemplating a reroof anyway, your best bang for the buck is to install a more reflective roofing system than dark shingles. Within any category (asphalt, metal, or tile), a lighter color can substantially reduce cooling loads with no price penalty. Compared with dark shingles, white-colored asphalt shingles give a modest reduction in attic heat and about a 4% reduction in space cooling. However, a highly reflective light-colored tile or metal roof can cut the heat flux through the roof dramatically. We have measured a greater than 30% reduction to the space cooling load during the hottest afternoons with reflective roofing systems.   — *Danny Parker*

Latent Cooling Loads

Q. I've heard installers of air-conditioning systems refer to "latent heat." What does this term mean?

A. When sizing an air-conditioning system, there are two types of cooling loads that must be evaluated: sensible load and latent load.

Sensible cooling load refers to the air temperature of the building. Factors that influence the sensible cooling load include sunlight striking windows, skylights, and glass doors; the insulation value in exterior walls, ceilings under attics, and floors over open crawlspaces; air infiltration through cracks in the building, doors, and windows; and (primarily in commercial work) the heat output of lights, appliances, and other equipment operated in the summer. All of these factors are included in an HVAC system designer's energy load calculation.

Latent cooling load refers to the energy required to remove humidity from the air. It takes more energy to condition humid air than to condition dry air.

It's sometimes a hard concept to wrap one's mind around because the latent load is not measured by an ordinary thermometer. Understanding the physics of water vapor may help.

While conduction, convection, and radiation are the heat transfer mechanisms of sensible heat, evaporation is a heat transfer of latent heat. When water evaporates, it changes phase by transforming from a liquid to a vapor. Latent heat is the energy required to overcome the molecular forces of attraction between the particles of a liquid. This energy allows the molecules to separate and the liquid becomes a vapor — where the attractions between molecules are minimal. When the material (in this case water) changes from a liquid to a vapor, there is no change in temperature between the liquid state and the vapor state. But while there is no sensible change in temperature, the air gains latent heat by gaining moisture.

HVAC system designers must account for the latent heat load — that is, the air-conditioning load required to remove humidity from the air. This latent load is a component of the energy required to maintain comfort.

From a system designer's point of view, this latent heat may need to be removed by condensation (the typical scenario in a hot, humid climate).

— *Clayton DeKorne*

One Furnace or Two?

Q. What factors must be considered when deciding whether to use a large hot-air furnace with two zones or two smaller furnaces?

A. A single unit is almost always more cost-effective than multiple smaller units. Multiple units should be considered only if the house design will not permit adequate means of ducting the various zones from a single heating plant.

With a single unit, the installed cost, operating costs, and life-cycle maintenance costs are almost always lower. If the customer wants options like air conditioning, a humidifier, or an air cleaner, the installed cost advantage in favor of a single unit is even greater. Zone system manufacturers claim that the operating cost of a single unit can be 30% lower than the cost of two smaller units. — *Jeri Donadee*

Dehumidifying the Air

Q. Can a dehumidifier be added to a forced-air system? If so, what should a contractor know about specifying this?

A. It is possible to add a dehumidifier to a forced-air system, but it's far simpler to use a stand-alone dehumidifier. Moisture vapor moves freely throughout the house, so humidity control in one place will work for the entire house. This means that a portable dehumidifier is usually enough to control humidity when you can't do so by eliminating the moisture source or by using an exhaust fan. (These should always be your first choices for controlling moisture.) A dehumidifier with a 40-pint capacity is sufficient for most houses.

In humid climates, it works best to place the humidifier and the return duct in the same vicinity. Sometimes, this is done by placing both in a central closet with louvered doors to allow ample airflow.

A dehumidifier is essentially an air conditioner that discharges the warm air back into the house rather than exhausting it outside. Therefore, an air conditioner can also be used to dehumidify a house. But since an air conditioner only dehumidifies when it is running, an oversized system will cool the house before the humidity is brought under control. If little cooling is desired, but you want to dehumidify the house, it's possible to size a central air conditioner to a quarter or half the size needed for normal cooling and run the system continuously without overcooling the house. However, the energy involved in doing this will be greater than that used by a portable dehumidifier.

In superinsulated houses with very low cooling loads, we often recommend installing a small (5,000 Btu) window air conditioner somewhere in the house to control humidity, and a central cooling system to control temperature. The window unit can be removed during the winter when dehumidification is not needed.   — *Henry Spies*

Ductwork

Water in Under-Slab Ducts

Q. Nine years ago I built a new house on a slab foundation. We put in a forced-air heating system, and the HVAC contractor ran the ducts under the slab. The house is on a lake, and during certain times of the year rising groundwater enters the ducts.

We installed a small pump with a float switch in the ductwork next to the furnace, and it has been doing a fairly decent job, but maintenance of the pump has been a continuing problem. The homeowner wants a more permanent solution. What is the best way to prevent water from entering the ducts?

A. Abandon the below-grade ductwork. Otherwise, the humidity in the ductwork will create a mold museum for the occupants, and they will be breathing a little of every water-soluble contaminant in the area. Plug the ducts at all penetrations through the slab with concrete. Route new ducts above the slab somewhere, or move to hydronic or resistance heating. No exceptions. *— Bill Rose*

Deteriorated Duct Board

Q. We are renovating a 1980 home. The HVAC sub says that since older fiberglass duct board disintegrates from the inside out, all of the ducts should be replaced. Could this be true?

A. When duct board in a crawlspace or attic is not sealed with UL 181 duct tape, condensed moisture can accumulate near unsealed duct joints, leading to duct deterioration. If such deterioration is severe, replacement of the ducts may be necessary. But it is unusual for properly installed duct board to "wear out."

Your subcontractor's diagnosis can be easily confirmed by a visual inspection. Simply cut open a section of duct and look at the interior surface to see if there is any flaking. If the duct is sound, be sure to reseal the duct using appropriate UL 181 mastic or tape. *— Jeri Donadee*

Radiant Heat Efficiency

Q. My HVAC sub tells me that hot water radiant heat is "more efficient" than traditional forced hot air. Is this true?

A. When applied to home heating systems, the term "efficiency" most often refers to the actual energy used, as reflected in the monthly utility bill. One of the main arguments used by those who claim that a radiant heating system uses less energy than a convected air system is that occupants will feel comfortable at a lower air temperature with radiant heat. Unfortunately, scientific studies don't always support this. Unofficial side-by-side comparison testing within the radiant heat industry has demonstrated utility savings, yet no such savings have ever been documented by a reputable study.

Exact energy-efficiency numbers are elusive because of the tremendous number of variables involved. Two houses can be built side by side with everything exactly alike except for the heating system. A comparison can then be made through the heating season that demonstrates the energy efficiency of one heating system versus the other. But the results, while interesting and informative, cannot then be applied generally to all installations (which is unfortunately too often done).

In the absence of any hard data, claims by radiant heating contractors that radiant heating systems provide energy savings must be seen as marketing puffery. — *Larry Drake*

Forced Air vs. Hydronic Heat

Q. What are the advantages of hot water heating systems vs. warm air systems?

A. Books have been written on this subject. In brief, both systems are capable of producing identical comfort levels if properly designed. In general, a pumped hot water system will cost more to install than a warm air system, although the use of plastic piping has decreased the difference. Furnaces and boilers are available in similar efficiencies, so operating costs are about the same for either system.

Where summer cooling is required, forced air systems make more sense. There is no reasonable way to use a hydronic system for cooling. You can install individual air handling units, which blow air over chilled water in the system. However, this can be expensive and noisy, and essentially converts the water system to an air system. If a separate duct system is required for cooling, the total cost of the separate heating and cooling systems far exceeds the cost of a combined air system.

The separate system installation does have some advantages, however. Hydronic heat can easily be supplied along outside walls under windows, which is the best place to deliver a heat source. And the cooled air can be supplied from registers in the center of the ceiling or high on the inside sidewalls, which is the best place to introduce cooled air. The duct system should have supply registers and return grilles that can be sealed during the heating season to prevent condensation in the ductwork when the system is not in use.

Air movement and dust control are other factors you should consider. There is less perceptible air movement with a hydronic system, since a hydronic heat system has no blower. This can be an advantage or a disadvantage. Many people prefer some noticeable air movement (which may account for the popularity of paddle fans), while others, particularly elderly clients, consider air movement to be a "draft." If an air system is properly designed, however — with the size and placement of supply registers carefully chosen — air movement should not be noticeable.

Heat pump systems are another matter, though. They move about twice as much air, and at a lower temperature, than furnace-supplied systems. Since this moving air might be uncomfortable, heat pumps are not recommended for use in housing for the elderly.

Dust control can be a problem with either system. Dirt streaks on walls are as common above hydronic baseboards as above warm air supply registers. Much work has been done in the hydronics industry to minimize streaking, but it still occurs with some designs. An air system can be equipped with filters that clean the air and reduce dust. The ordinary glass fiber furnace filter removes only about 5% of the dust from the air passing through it. High-efficiency mechanical filters, using a pleated paper filter element, remove about 95% of the dust in the air stream. Electronic filters remove about 99%, including smoke particles. Activated charcoal filters can be added to remove odors and organic chemicals.   — *Henry Spies*

Sizing a Steam Boiler

Q. I am remodeling a large older home that is heated by a low-pressure steam system. Because of the size of the house, steam seems to be a good system to use, but the boiler is obsolete. We are adding insulation and new double-glazed windows. How can I calculate the required size of the new boiler? Or can the system be converted to hot water?

A. In a steam system, the boiler is not sized according to the calculated heat loss of the building. Instead, it must be sized according to the number of square feet of installed radiation. Tables that give the equivalent square feet of radiation are available for most old radiators from the Hydronics Institute Division of Gas Appliance Manufacturers Association (908/464-8200, www.gamanet.org). A new radiator should include that size in the specs. Steam boilers normally carry a specification plate that gives the number of square feet of radiation the boiler can supply.

If the boiler is sized according to the calculated heat load, it will not supply enough steam to reach the far end of the system without condensing first, and the system will never work properly. This is a common problem when conversion burners (gas or oil) are installed in old coal-fired boilers. The burner must be large enough to generate the same amount of steam as the coal-fired unit, and that is often two or three times as large as the burner size derived from the calculated heat loss. A steam system that is underfired can have an efficiency approaching zero, regardless of the reported efficiency of the boiler.

If the house has a two-pipe steam system, as opposed to a one-pipe system, it can be converted to a pumped hot water (hydronic) system with minor changes in the piping (removing the steam traps and the Hartford loop). You can then use the same radiators. The hydronic systems will produce a more even heat, especially if the system temperature is controlled to match outdoor temperatures. By contrast, a steam radiator is either on or off. For more information on converting steam systems to hydronic, contact the Hydronics Institute Division of GAMA.   — *Henry Spies*

Non-Asbestos Pipe Wrap

Q. What material can be used to reinsulate steam pipes after the old asbestos insulation is removed?

A. The most common insulation now used on steam pipes is molded fiberglass sleeve with a plastic or fabric cover. This material is widely available from plumbing and heating supply houses. The fiberglass sleeves come in precut lengths for different pipe diameters (measured as inside diameters), and usually have self-sealing tape attached to seal the cover.   — *Henry Spies*

Combining Radiant and Baseboard Heat

Q. I have a hot water system feeding high-temperature baseboard heaters and a low-temperature radiant slab. How should the system be set up to accommodate these separate zones?

A. There are several ways of combining low-temperature radiant-floor heating with higher-temperature distribution systems, such as fin-tube baseboard convectors. Above all, the system must be designed:
- to prevent condensation within the boiler, and
- with a specific type of control system.

Conventional boilers must have return water temperatures high enough to prevent sustained condensation on the fire side of the boiler, or within the flue pipe. Water vapor is a by-product of combustion, and if allowed to condense, it can cause severe corrosion. Flue pipes are especially vulnerable, and can fail in a matter of weeks when condensation is present. This could allow toxic gases to be released into the building.

Typically, the return water temperature for a gas- or oil-fired boiler should be 140°F or higher to prevent condensation. Since radiant systems operate with a return water temperature in the range of 80° to 100°F, their return water must be mixed with hotter water before it is sent back to the boiler. There are two simple ways to do this — with a four-way valve or with injection mixing.

The four-way mixing valve lowers the temperature of the water supplied to the radiant floor system by mixing return water into the radiant loop, as shown in the illustration (Option 1). To avoid condensation in the boiler, the four-way mixing valve maintains a relatively high return

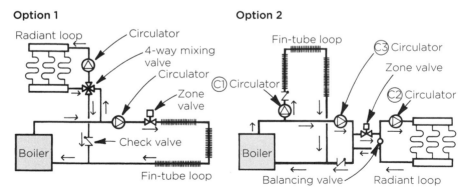

When combining radiant and baseboard heat, you must prevent condensation in the boiler caused by the low temperature of the radiant heat's return water. One method uses a four-way mixing valve (Option 1) to mix hot supply water into the radiant heat's return water. Another approach (Option 2) uses a zone valve to deliver hot water to the radiant loop and a circulator (C3) to heat the return water.

water temperature by mixing some hot supply water into its return flow.

To control the four-way valve, you'd ideally have a motor-operator regulated by an outdoor reset control. This measures outdoor temperatures and automatically adjusts the valve to maintain a suitable water temperature in the radiant floor to match the required heating load. A less expensive (and less exact) control system for the four-way valve is to control the circulator in the floor heating loop with a room thermostat. This way, the four-way valve is set at the design-load temperature of the floor heating system and left there. When heat is needed, the circulator comes on to deliver hot water to the area. The thermostat should have a low differential (one or two degrees) in order to minimize swings in room temperature.

Another way to lower the water temperature of the radiant floor is through injection mixing (see illustration, Option 2, previous page). In this system, the high-temperature zones have individual circulators (C1) that are controlled by room thermostats, just as in a standard multizone system. Water circulates continuously (using circulator C2) through the radiant floor loop during the heating season, and a zone valve opens to allow hot water to flow into the loop when heat is needed. This zone valve can be controlled by a thermostat, or for more precise control, by a reset control.

The hot water from the zone valve is mixed with cool return water at the tee downstream from the valve. A balancing valve determines how much hot water flows into the radiant loop when the zone valve is open. To prevent condensation, the circulator in the main system loop (C3) must operate when the zone valve is open to shunt a significant portion of hot water back toward the boiler.

The injection mixing hardware (zone valve plus reset control) is significantly less expensive than the four-way valve system with a reset control (about $300 vs. $800). Detailed information about both approaches can be obtained from Tekmar Control Systems (250/545-7749, www.tekmarcontrols.com).

— John Siegenthaler

Heating Tubes in Basement Slabs

Q. A client wants a portion of a basement slab to have radiant heat to take the chill off the floor of a planned playroom (there will be supplemental heat). The HVAC contractor wants the radiant tubing, which will be attached to wire mesh, to be lifted into the middle of the slab during the pour to put the heat closer to the surface. The concrete contractor doesn't want to do this because he insists that cracks will show up along the tubing. He wants to leave the tubing at the bottom, and he says the insulation board will drive the heat up anyway. He recommends at least 3 inches of concrete above the tubing. Which is correct?

A. We typically try to keep at least an inch of concrete above our tube. If the slab is 4 inches thick and the mesh is in the middle, a 5/8-inch o.d. tube would have about 1 1/4 inches above it. Pulling mesh up during the pour is by nature somewhat inaccurate, so it's probably best to err on the side of keeping the mesh low in the slab. I don't believe the height of the tube in the slab is going to make a significant difference in system performance even if it's on the bottom — the entire slab will be heated in any case. My advice is to let the concrete contractor have it his way.

One point about laying out the tube: Tie it perpendicular to the wire as much as possible, away from parallel strands of wire. The reinforcing value of the wire depends on concrete bonding all the way around it, so it's best to keep the tube out of the way.   — *Bill Clinton*

Too Much Thermal Mass?

Q. When designing a radiant floor, can there be such a thing as too much thermal mass? Here in Alaska, we sometimes see the temperature jump from -10°F up to 40°F above in just a few hours.

A. Thermal mass can be a double-edged sword. With too little mass, the boiler runs more often, and there is an increased likelihood of indoor temperature swings. With too much mass, the system can be slow in responding to sudden outdoor temperature changes, resulting in spikes and valleys in the indoor temperature.

In the case of high-mass systems, there are ways to reduce this effect. Outdoor temperature controls, which are common in Europe, can be used in what is called a "constant flow" system. With this method, also called "flat lining," the temperature of the fluid is modulated based on the outdoor air temperature. If the heat loss calculations are accurate, you won't notice any variation at all in a building's indoor temperature, no matter how quickly the outdoor temperature changes.   — *Doug Mossbrook*

Using Dry Sand in a Radiant Floor

Q. How do you go about using sand instead of concrete for thermal mass in a radiant floor?

A. Avoid using sand that is too wet. A pile of sand stored outdoors can be quite wet in the middle. I recommend buying play sand (sand sold for use in children's sandboxes), which comes in a sealed bag. Play sand has a dependable, reasonable moisture content.

To prevent the sand from escaping through the subfloor, you need to be able to seal the cavity well. Either install polyethylene over the sub-floor, or (if the plywood has tongue-and-groove joints) you can caulk the edges of the plywood. Install the sleepers over the floor, and then lay in the sand and screed it level with the tops of the sleepers.

To make the sand easier to work with, you may want to mix in a small amount of Portland cement — just enough to firm it up.

Next, turn on the heat to let the sand dry. Once the sand is dry, install a second layer of polyethylene to keep the sand in place. You can place hardwood flooring directly on top of the polyethylene, or you can use plywood underlayment. Remember that plywood is an insulator and reduces heat transfer somewhat. — *Doug Mossbrook*

Hickory Flooring Over Radiant Heat

Q. A customer wants me to install hickory flooring over radiant heat. Does hickory present any unusual problems in this application, compared with other species?

A. Hickory, which is a little more dimensionally stable than red oak, does not present any unusual problems over radiant heat. As with any installation of solid wood flooring over radiant heat, the National Wood Flooring Association recommends that only narrow boards ($3^1/4$ inches wide or less) be used. For additional dimensional stability, it is preferable for the wood to be quartersawn or rift-sawn. — *Bonnie Holmes*

Will Carpet Stifle a Radiant Slab?

Q. I'm finishing a basement room for customers who want to put carpet over the radiant slab. I'm concerned the carpet will insulate the slab, reducing heat transfer, but I can't talk them out of it. The slab is 4 to 5 inches thick, and the 1/2-inch PEX tubing is laid out on 12-inch centers, with 1 1/2 inches of rigid foam insulation under the slab and around its edges. The slab is at least 5 feet below grade; local frost depth is 48 inches. Is there a type of carpet and pad that would allow more heat to radiate into the room? Will the carpet cause heat to be lost into the ground?

A. There are many successful radiant heating installations in which carpet is placed over a concrete slab. For best performance, use a low-pile, commercial-grade, level-loop carpet bonded directly to the top of the slab, which will provide low thermal resistance to upward heat flow. If a pad must be used, it should be a low-resistance slab rubber pad approximately 1/4 inch thick, which will add about 0.31 to the upward R-value of the carpet (avoid polyurethane pads because of their higher R-value). Given the tube spacing you have, and the fact that basement heating loads are typically low, adding the slab rubber pad will likely raise the required circuit water temperature about 5°F.   — *John Siegenthaler*

Thin Radiant Slab

Q. I want to install a thin radiant slab over a wood subfloor in a remodel. The flooring will be ceramic tile. What is the thinnest slab I can get away with?

A. If you want a slab as thin as possible and expect to tile over it, I think regular concrete is out of the question. You're bound to get cracking, which will telegraph through the tile. My choice would be a gypsum-cement product — either Gyp-Crete from Maxxon Industries (800/356-7887, www.maxxon.com) or Gyp-Span from Hacker Industries (800/642-3455, www.hackerindustries.com). Gypsum cement, unlike concrete, doesn't shrink as it cures and usually won't crack.

When it comes to preventing cracks, the most important factor is the rigidity of the floor framing and subfloor.

Gypsum cement can, with care, be poured as thin as 1 1/4 inches. You can thin-set the tile to a gypsum-cement slab. But before tiling, you need to allow the gypsum underlayment to dry completely, and then seal it. In bathrooms, you will need a waterproof membrane, such as Noble Company's NobleSeal (800/678-6625; www.noblecompany.com).   — *Bill Clinton*

Fixing a Radiant-Tubing Leak

Q. I have a leak in my recently installed radiant tubing, no doubt caused by freezing last winter before the system was fully operational. The tubing is a 300-foot run buried in a 4-inch concrete slab. Is there some kind of "stop leak" that can be circulated in the system to plug the leak? Is there a way to locate the leak?

A. As far as I know, there's no product that can be circulated through any type of radiant tubing that will stop a leak. So you'll have to cut out the damaged tubing and splice in a new section.

First, though, you have to find the leak. The only reliable method I know of is to run warm water through the system and view the slab with a thermal imaging camera; the leak will appear as a plume spreading out from the tubing. This will work best if the slab is relatively cool when the test is initiated.

Once you've located the leak, mark the location on the slab, chip away the concrete to expose the tubing, and patch the leak. Check with the tubing manufacturer for the required fittings and procedure to do the repair, and be sure to pressure-test the circuit before patching the slab.

The hardest part of this process may be locating a thermal imaging camera. Unless you have access to one owned by a local utility, volunteer fire department, or other agency, you'll need to contract with an infrared inspection service. — *John Siegenthaler*

Solar Heat

Cost-Effectiveness of Solar Equipment

Q. How cost-effective are solar panels?

A. The term "solar panel" can refer to two different types of equipment: photovoltaic (PV) modules or solar thermal collectors. PV modules generate DC power to charge batteries, or when connected to an inverter, to power house loads or supply electricity to the utility grid. Solar thermal collectors are used to heat swimming pools or to supply a portion of a home's domestic hot water or space heating needs.

Cost-effectiveness calculations are complicated and site specific. In sunny parts of the country, like Arizona or New Mexico, a PV module or solar thermal collector may have two or three times the annual energy production of the same equipment installed in northern Vermont.

Cost-effectiveness calculations also depend on the cost of available conventional energy; when replacing natural gas, a solar hot water system will have a much longer payback than when replacing electricity. Moreover, rising energy prices can quickly change the cost-effectiveness of a proposal from marginal to reasonable.

Finally, some utilities are now offering rebates to cover up to half the cost of PV or solar thermal equipment; such incentives obviously improve cost effectiveness.

Here's the summary: PV systems are not yet cost-effective anywhere in the country where utility grid power is available. Where significant rebates are available, however, a PV system installed on a new home in a sunny climate can produce a positive cash flow for the homeowner if the cost of the system is included in a new home mortgage.

Solar thermal collectors are a cost-effective way to heat swimming pools in all areas of the country. In areas with lots of sun and high electric rates, like Hawaii, solar domestic hot water systems are very cost-effective. In California, Florida, Arizona, and New Mexico, and other areas with similar climates, solar hot water systems are also cost-effective, especially if they are being substituted for an electric water heater, although the payback period will be longer than in Hawaii.

In most northern states, if maintenance costs are included, an investment in a solar hot water system is probably a break-even proposition, especially when replacing natural gas. Of course, any increase in energy prices will improve the cost effectiveness of all types of solar hardware.

— *Martin Holladay*

Solar Shingles

Q. Solar modules that can be installed in place of roof shingles seem to make more sense than PV modules mounted on racks above the roof. Besides being much less noticeable, it seems that using photovoltaic shingles would save on the expense of installing a regular roof. Why aren't more people using them?

A. Actually, many people are using these solar modules — we call them building-integrated photovoltaics, or BIPVs — as roofing. Designed to replace long-lived roofing systems such as composite slate shingles or concrete tiles, or to be installed over standing-seam metal roofing, BIPVs are normally installed by a roofer as part of a new roof. Because most of these roofs are actually a mixture of PV modules and conventional roofing material, the roofer has to weave the two together, while a solar contractor makes the necessary electrical connections and supervises the work.

There are several reasons why more BIPV systems are not being installed. For one thing, most PV systems being installed today are for retrofit, in cases where a new roof isn't required. Second, of those home-owners who do need a new roof, many choose composition shingles, and the only current manufacturer of PV equipment designed to integrate with composition shingles has had trouble keeping up with the current demand. Third, an integrated PV roof is not necessarily less expensive than putting modules above the roof; often, it's actually much more cost-ly due to higher material costs, increased wiring costs, and smaller mod-ules that require more total labor to install.

Fourth, long-term maintenance and repair of a BIPV system may involve removing and replacing the roof itself, a prospect that concerns some homeowners (even though the BIPV usually has a 25-year warran-ty). Finally, most solar companies would prefer to install their systems themselves rather than deal with the logistics and costs of hiring, train-ing, and supervising a roofer.

BIPV makes the most sense for new construction, especially when the builder can hide the system from view in plain sight. As more new-home builders and developers get wise to the advantages of offering PV to their customers, expect to see many more of these BIPV roofs dotting the landscape (if we can spot them).   — *Gary Gerber*

Plastic Pressure-Relief Piping

Q. Is it acceptable to use plastic piping for the pressure-relief discharge on a domestic water heater? The code in my area requires that the discharge piping be rated at or above the temperature of the system, but it's not clear whether that refers to the temperature of the water in the tank or the temperature setting of the relief valve.

A. Without protection, a domestic water heater whose thermostat has failed would see a continuous rise in temperature and pressure. When the water pressure exceeded the capacity of the tank (typically 300 psi), the tank would burst with enough force to send an average-size car 125 feet in the air.

A temperature- and pressure-relief valve is designed to prevent such catastrophic failures. In my area, we follow the IAPMO (International Association of Plumbing and Mechanical Officials) Uniform Plumbing Code, which allows the use of galvanized steel, hard-drawn copper, chlorinated polyvinylchloride (CPVC), polybutylene (PB), or code-listed straight sections of relief-valve drain tube.

It's true that the temperature ratings for CPVC and PB are well under the settings (210°F or less) for most relief valves. The consensus among inspectors I've talked to, however, is that the relief cycle for an excessively high-temperature or high-pressure condition is so intermittent and short-term that CPVC or PB are acceptable materials for this application. Ultimately, however, it's up to the local inspector to decide, so check with local code enforcement first. — *Redwood Kardon*

Water Heater Relief Valve Keeps Tripping

Q. I have decided to replace my boiler with a 75-gallon propane water heater. In general, this is working great, except that about every three weeks the pressure-temperature relief valve, which is rated at 210°F, keeps tripping. I live in the cold climate of Michigan, and I need to keep the water heater temperature set on

"High," which is about 180° to 190°F. When the relief valve trips, I notice that the water temperature is about 195°F. What's going on?

A. Although most residential water heaters have a maximum setting of 160°F, there are some heaters that are factory-equipped with 180°F thermostats. I have used a number of these without having the problem you describe.

When the burner on a water heater turns off and there is no circulation happening, hotter water can stratify at the top of the tank, resulting in higher temperatures than the thermostat setting. Perhaps this is the cause of your readings. If you are sure that your temperature reading is accurate, I would turn down the setting a bit and try installing a new relief valve. Do not, under any circumstances, attempt to operate a water heater without a proper relief valve.

What is likely is that you have a pressure problem and not a temperature problem. Thermal expansion may be forcing the relief valve open. Do you have an expansion tank installed? If so, is it big enough? Was its air charge adjusted to equal the household pressure before installing it? A pressure gauge attached to the drain of the water heater will help you determine if your problem is really excess pressure.

Finally, remember that temperatures this high can be dangerous. Unless you are well-trained and quite competent, don't risk working with that water heater when it's hot. — *Bill Clinton*

Using an Electric Water Heater for Radiant Heat

Q. Will an electric water heater work in a radiant system to heat a house? How about a tankless electric water heater?

A. Yes, an electric water heater will work fine. Of course, in most parts of the country, electricity is relatively expensive to use for home heating. However, in areas of the Pacific Northwest where electricity is fairly cheap, using an electric water heater can make sense and will make for a very simple reliable system.

I would, however, stay away from the instantaneous electric heaters. They operate with fairly high pressure differentials, which makes pump selection difficult. You could also have problems if the electric input were substantially higher than the load, since this would cause excessive cycling and perhaps excessive temperatures. The advantage of a system with a storage tank is that the tank helps buffer the system from temperature and pressure extremes. — *Bill Clinton*

Chapter 7: HVAC/*Water Heaters*

Energy Payback From Instantaneous Water Heaters

Q. Do the energy savings provided by instantaneous gas water heaters justify their higher cost?

A. Before you consider installing an instantaneous gas water heater for energy efficiency, consider whether your customers will be satisfied with the flow rate of the model chosen. The most common models of instantaneous water heaters have maximum flow rates in the range of 2 to 3 gallons per minute. Three gpm is the bare minimum to supply two simultaneous showers, and most American families expect their water heater to provide up to 6 gpm of hot water when necessary.

Assuming you've decided to install an instantaneous gas water heater with at least 3 gpm of flow, a typical choice would be the AquaStar 240FX (800/642-3199, www.boschhotwater.com), which is available for about $900 to $1,030, depending on whether you need the outside vent hood. The AquaStar 240FX has an efficiency factor (EF) of 0.84. An ordinary 40-gallon gas water heater with an EF of 0.56 costs about $270, so you would need to save around $700 on your fuel bill before your energy savings would repay the added cost of the instantaneous heater. An instantaneous water heater will probably last 10 to 20 years, compared with 7 to 10 years for a conventional gas water heater.

Investing in an instantaneous model makes the most sense for those with high fuel costs. (See the table, which is based on hot water use of 64.3 gallons per day, or 23,470 gallons per year, by the "average U.S. household," as shown in the Energy Guide Labels.) If you have access to natural gas, you won't save much, since the payback period (assuming that gas costs $1.30 per ccf) is about seven years. If you're burning propane, however, the payback period is shorter: at $2.60 a gallon, it would be about three years.

Of course, many other factors can affect payback calculations, including how much hot water is used (high-use families see a quicker payback), differences in maintenance costs, the likelihood that an

	Purchase price of water heater	Annual gas consumption (in therms)	Annual natural gas bill (at $1.30 per ccf)	Annual propane bill (at $2.60 per gallon)
Conventional gas water heater (EF 0.56)	$270	267	$348	$760
Instantaneous gas water heater (EF 0.84)	$970	187	$244	$511

instantaneous heater will not need to be replaced as frequently as a conventional heater, and possible differences in installation costs (an instantaneous heater may require a larger gas supply line and a larger flue than a conventional water heater). — *Martin Holladay*

Fireplaces & Chimneys

Makeup Air for Fireplaces and Exhaust Fans

Q. In a house I am building, I need to provide adequate makeup air for a fireplace and a 600-cfm cooktop exhaust fan. How do I size a passive duct to introduce exterior makeup air into the house?

A. A fireplace and a large cooktop exhaust don't belong together in a modern house. To understand why, let's examine the issues:

Whether or not the fireplace or exhaust fan will cause problems depends on several factors. The most important factors are the leakage characteristics of the house, the type of combustion devices used for space and water heating, the location and type of fireplace and chimney, and the presence of other exhaust devices such as a clothes dryer, central vacuum, or bath fans.

The quantity of combustion and dilution air needed by a fireplace varies, depending on the type of fireplace, the type of chimney, and the size and burning phase of the fire. The amount might be as little as 100 or 200 cubic feet per minute (cfm) at the start-up or burn-down of a small fire in a fireplace with glass doors, or as much as 800 cfm or more for a large fire in an open fireplace.

If the house is very leaky, drawing that much air may cause only a small negative pressure. But if the house is tighter, pulling that much air may cause a very large negative pressure. Although negative pressure can interfere with the combustion of the fireplace, in many cases it simply limits the magnitude of the fire, somewhat like a woodburning stove with good air control.

Negative pressure can certainly interfere with atmospherically vented furnaces, boilers, or water heaters, potentially causing spillage of flue gases. That's why you would be well advised to choose sealed-combustion appliances. Also, excessive soil gas and/or garage gases could be drawn into the home when the fireplace is in use.

Let's assume for a moment that you want to limit the negative pressure caused by the fireplace to 5 pascals. The first table shows the size of combustion air opening needed, assuming three levels of combustion and dilution air requirements (see Table A, next page).

If your house is very leaky, existing leaks may provide adequate combustion and dilution air for your fireplace. If your house is very tight, you will need a very large passive air opening.

Now let's focus on the 600-cfm cooktop exhaust fan. If we know the

A. Makeup Air Required for a Fireplace

House Tightness	Blower Door (cfm @ 50 Pa)	Combustion Air Opening for:		
		200 cfm	500 cfm	800 cfm
Very Tight	600	28 sq. in.	154 sq. in.	280 sq. in.
Pretty Tight	1,200	None	98 sq. in.	224 sq. in.
Kind of Tight	1,800	None	41 sq. in.	168 sq. in.
Typical	2,400	None	None	111 sq. in.
Loose	3,000	None	None	54 sq. in.
Very Loose	3,600	None	None	None

B. Negative Pressure Caused by a 600-cfm Exhaust Fan

House Tightness	Blower Door (cfm @ 50 Pa)	Predicted Negative Pressure Due to 600-cfm Exhaust Fan
Very Tight	600	50 pascals
Pretty Tight	1,200	17 pascals
Kind of Tight	1,800	9 pascals
Typical	2,400	6 pascals
Loose	3,000	4 pascals
Very Loose	3,600	3 pascals

C. Makeup Air Required for a 600-cfm Exhaust Fan

House Tightness	Blower Door (cfm @ 50 Pa)	Makeup Air Duct Required
Very Tight	600	17-inch diameter
Pretty Tight	1,200	16-inch diameter
Kind of Tight	1,800	14-inch diameter
Typical	2,400	12-inch diameter
Loose	3,000	10-inch diameter
Very Loose	3,600	5-inch diameter

cfm @ 50 pascals of the home using a blower door, we can easily predict the negative pressure due to the operation of this fan (Table B).

A negative pressure of 3 or 4 pascals — which this exhaust fan can cause, even in a very leaky house — can pull the products of combustion down a fireplace chimney. This is especially likely during the start-up or burn-down phases of a fire. The simplest solution to this problem is to never use the cooktop exhaust fan when the fireplace is being used.

However, remember that this same negative pressure could cause spillage of combustion gases from an atmospherically vented space or from water-heating equipment. With this type of exhaust device, sealed combustion equipment would be advised for space and water heating.

If you decide that these approaches are impractical, you can try to size a makeup air inlet for the cooktop exhaust fan. Let's assume a 20-foot smooth duct with three 90-degree elbows and a screened hood. Let's also assume that the fireplace (or water heater) can tolerate 3 pascals. Assuming no house leakage, you would need an opening of approximately 325 square inches, or the equivalent of a 20-inch-diameter duct. If we know the house leakage as measured by a blower door, we could use Table C to size the makeup air opening.

If the cooktop exhaust fan will be used while the fireplace is in operation or with atmospherically vented space and water heating, you should install a makeup air fan.

Now, if that isn't complicated enough, there is one last nagging issue. When the cooktop exhaust fan is used when the fireplace is not, the makeup air often comes down the fireplace chimney, commonly leading to homeowner complaints about a sooty odor in the house. Unfortunately, it is very difficult to seal the chimney completely.

In conclusion, a fireplace and a large cooktop exhaust fan just don't belong together in the modern house. The fireplace by itself can be handled by selecting sealed combustion equipment for space and water heating and introducing the necessary combustion and dilution air, especially for tighter homes and during the critical start-up and burn-down phases. The cooktop exhaust fan can likewise be handled by itself with proper selection of combustion equipment and some provision for makeup air. It is very difficult or impossible, though, to handle combustion and makeup air requirements when a house has both a fireplace and a cooktop exhaust fan. If both are inextricable parts of your plan, I would recommend a different type of hearth product and a kitchen cooktop or exhaust fan with much smaller cfm requirements.   — *Pat Huelman*

Installing a Metal Fireplace

Q. When installing a zero-clearance metal fireplace, are there any requirements for fire stops or draft stops? Does the inside of the chase need to be drywalled?

A. If there is one situation where you need to follow the manufacturer's installation instructions, it is with metal fireplaces. They are sold as a system, with brand-specific chimneys. The components required vary from manufacturer to manufacturer.

When the chimney is enclosed in a chase, some manufacturers require that the chase be equipped with two vents, one at the floor level and one near the ceiling, to help dissipate heat inside the chase. You need to install fire stops at floor penetrations where specified. Because the outside diameters of metal chimneys vary, the fire stops are manufactured components specific to the type of fireplace you are installing.

Although each system varies, in many cases the installation instructions will specify requirements for wall bands to support the chimney and attic insulation shields where applicable. Instructions should include a chart for calculating lateral distances when installing offsets and offset supports in the chimney.

Get a copy of the installation instructions before you start. This will give you the required clearances for the system, as well as information on what accessory components to order. — *Stephen Bushway*

Covering Up a Brick Fireplace

Q. I am building an addition that will enclose the back side of an existing brick fireplace, which is now on the exterior of the building. Can 2x4 furring be attached directly to the brick?

A. Unless the thickness of the chimney wall is at least 8 inches, most codes prohibit combustibles in direct contact with a chimney. The only exceptions are for pieces of wall trim and roof sheathing.

In the case of your addition, you will be creating an interior fireplace. In the area directly behind and around the firebox, all combustibles must be kept at least 2 inches away from the outside of the brick. Because of this requirement, in most cases the floor framing of the addition must also be kept at least 2 inches away from the chimney.

This would be the case when the fireplace hearth is at the same level as the new floor, although not when the hearth is substantially higher than the floor level.

You can build a stud wall around the chimney. If the studs are wood, there must be a 2-inch space between the studs and the bricks. Because steel studs are noncombustible, they can be installed against or attached to the bricks, as long as there are no combustibles, including drywall, within 2 inches of the chimney. Instead of framing out around the masonry, one simple solution is to parge the chimney with a coat of stucco, or to screw cement backerboard directly to the chimney. The cement board can be finished with a skim coat of drywall compound. If stucco or cement board is installed on the chimney, the surface can be painted but should never be wallpapered.

Finally, remember that your local code may have more stringent requirements than the 2-inch clearance required by the IRC, so check your local code. — *Stephen Bushway*

Cold Chimney Cure

Q. For a brick fireplace and chimney on a gable end wall, how should the joint between the framing and the brick be sealed? Can a thermal break be included in the masonry to stop heat loss?

A. If the chimney and fireplace are solid masonry, the usual procedure to seal the joint is to bed a trim board in a good elastomeric caulk against the brick and the sheathing.

I do not know of any material that would provide a thermal break and that has the required fire resistance and bearing capacity to be included in a solid masonry fireplace. It is possible to insulate a brick-faced fireplace by filling the cavity between the outer wythe of masonry and the flue with perlite or vermiculite (see illustration). A chimney-top damper can then be used to keep cold air from flowing down the flue. Of course, if thermal efficiency is a concern, it is always better to locate a masonry fireplace in the center of a house rather than at a gable end wall.

The difficulty in insulating a chimney chase and firebox is one reason prefabricated metal fireplaces, which can be installed in an insulated frame wall, are so popular. — *Henry Spies*

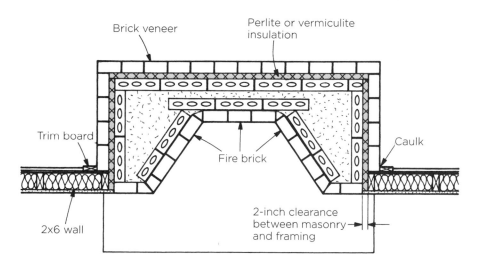

A solid masonry fireplace can create a thermal short circuit through a wall. This can be prevented by building a cavity wall filled with perlite or vermiculite (shown in gray).

Reducing Clearance to Woodstove Pipe

Q. Can I use Type X "fire code" gypsum board to reduce clearances to combustibles for a woodstove pipe? For example, could I attach a layer or two of gypsum board to the side or bottom of a nearby cabinet to reduce the required 18-inch clearance?

A. No, you can't use gypsum board to reduce clearances from combustion appliances and vent pipes. You may be thinking of the use of drywall in firewalls and other fire-resistive assemblies, which are intended to slow the spread of fire, not to prevent initial combustion.

You'll need a heat shield to reduce clearances to a woodstove pipe. One of the most common is a piece of 24-gauge sheetmetal, attached to the nearby wall or ceiling with standoffs so that there's a 1-inch air space behind it (see illustration).

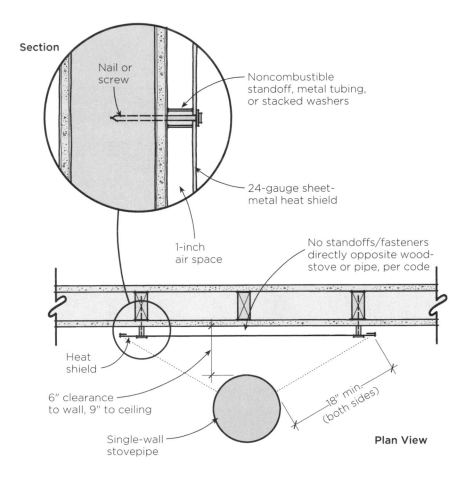

Section

Nail or screw

Noncombustible standoff, metal tubing, or stacked washers

24-gauge sheet-metal heat shield

1-inch air space

No standoffs/fasteners directly opposite wood-stove or pipe, per code

Heat shield

6" clearance to wall, 9" to ceiling

18" min. (both sides)

Single-wall stovepipe

Plan View

If properly installed, this can reduce an 18-inch single-wall pipe clearance to 9 inches overhead and 6 inches on the sides and rear, according to the IRC. You can also add a protector shield to the pipe itself.

Another option would be to upgrade to a double-wall pipe. For example, Simpson Dura-Vent (800/835-4429, www.duravent.com) makes a double-wall stainless-lined pipe that is rated for 6 inches to combustibles on the sides and 8 inches above. This is available in a black finish and might look better than a heat shield.

Most codes include a chart (see IRC Table M1306.2, for example) that lists several options in addition to the sheetmetal shield, but none of the other options is as simple. Check your local code for specifics, and make sure you look at the listed clearances for your woodstove. To have an approved installation, you've got to meet those requirements, too.

— *Don Jackson*

Boxing in a Metal Chimney Above the Roofline

Q. My company has been asked to enclose a triple-wall metal chimney with a rectangular chase above the roofline. The chimney vents a fireplace and extends about 4 feet above the roof. The homeowners want to box in the chimney with framing, sheathing, siding, and trim.

What type of clearance is necessary between the metal chimney and the box? What is the best way to flash the top of the box?

A. First, check the chimney manufacturer's printed specs to determine the minimum clearance between the chimney and the framing. In most cases, the minimum clearance will be 1 or 2 inches; if uncertain, err on the conservative side.

Frame a wooden box to match roof pitch. I usually build two side walls from 2x3s or 2x4s, and just use sheathing to tie the walls together at the up-roof and down-roof sides. Make the overall height of the box 6 inches short of the pipe termination exclusive of the weather cap. This leaves enough of the pipe exposed to install the weather collar and the cap. Assuming shingle or clapboard siding, install a 1x2 furring strip around the perimeter of the box flush with the top edge.

Fasten the box to the roof framing, preferably with screws or lag bolts. Install blocking between the roof trusses or rafters if necessary, so the screws bite into something. Check to be sure the box is plumb and square at the top. If necessary, frame in a cricket on the up-roof side. Install step flashing and roof shingles as usual.

Install the siding, and then cover the 1x2 furring at the top with a

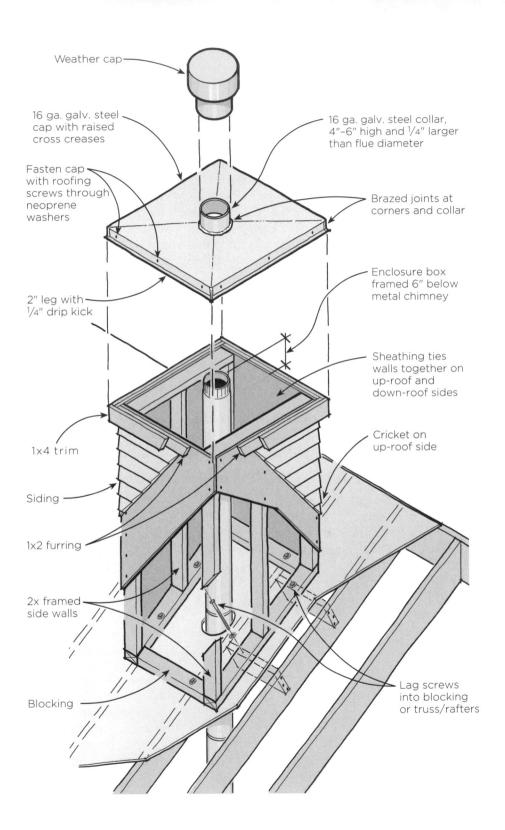

Weather cap

16 ga. galv. steel cap with raised cross creases

Fasten cap with roofing screws through neoprene washers

2" leg with 1/4" drip kick

16 ga. galv. steel collar, 4"–6" high and 1/4" larger than flue diameter

Brazed joints at corners and collar

Enclosure box framed 6" below metal chimney

Sheathing ties walls together on up-roof and down-roof sides

1x4 trim

Cricket on up-roof side

Siding

1x2 furring

2x framed side walls

Lag screws into blocking or truss/rafters

Blocking

1x3 or 1x4 piece of finish trim all around the box. Measure the top of the trimmed box, taking dimensions carefully. Make a sketch locating the chimney within the perimeter. Note the actual diameter of the chimney: it should be 1/4 inch larger than the pipe.

Make or have made a galvanized sheetmetal cap with a separate 4- to 6-inch-tall collar that slips over the chimney. I prefer to use 16-gauge material because it's more rigid than 20-gauge.

Diagonally crease the metal cap in both directions before installing the collar. The creases add rigidity to the cap and give a hip effect to shed water. The edges of the cap should fold over the trim about 2 inches, with a 1/4-inch drip kick along the bottom edge. Snip and bend tabs from either the top or the neck for a secure connection. Although some builders secure the tabs with pop-rivets and seal the seams with high-temperature silicone, I am doubtful that a caulked joint will be durable. It's best to braze or solder the joints in the galvanized metal. — *Mike Guertin*

Interiors

8

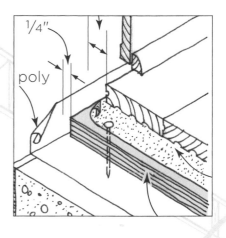

1/4"

poly

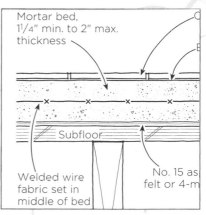

Mortar bed,
1 1/4" min. to 2" max.
thickness

Subfloor

Welded wire
fabric set in
middle of bed

No. 15 as
felt or 4-m

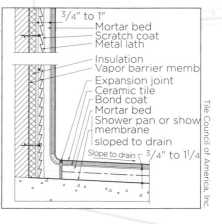

3/4" to 1"
Mortar bed
Scratch coat
Metal lath
Insulation
Vapor barrier memb
Expansion joint
Ceramic tile
Bond coat
Mortar bed
Shower pan or show
membrane
sloped to drain
Slope to drain 3/4" to 1 1/4

Tile Council of America, Inc.

Laying a Hardwood Floor Parallel to the Joists

Q. For aesthetic reasons, I would like to run $3/4$x3-inch oak strip flooring parallel to the floor joists, instead of perpendicular as is typical. Is there any reason I can't do it? The subfloor is $3/4$-inch tongue-and-groove plywood.

A. If the subfloor is stiff enough, there is no reason why solid nail-down strip flooring can't be run parallel to the joists. First, renail the existing plywood subfloor, so that the old nails are tight, and add new nails to achieve a maximum nail spacing of 6 inches.

Whether or not a $3/4$-inch plywood subfloor is adequate where floor joists are spaced 16 inches on-center is a judgment call. Some plywood subfloors deflect more than others. If the plywood feels stiff, you will probably be okay. If you can feel some deflection, you have two options. If raising the floor height is not a problem, you can install a layer of $1/2$-inch plywood over the existing $3/4$-inch plywood. The other option — assuming you have access to the open joists from below — is to install 2x4 blocking between the joists, 24 inches on-center. Once the blocking is toe-nailed in place, it should also be fastened from above with screws through the plywood.

In the unlikely event that the existing floor joists are spaced

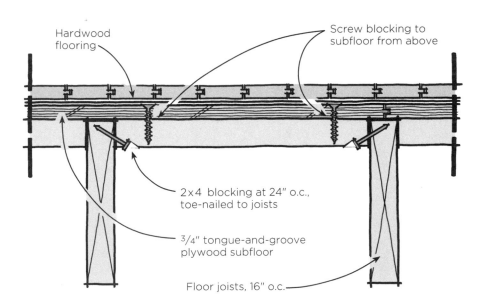

Hardwood flooring

Screw blocking to subfloor from above

2x4 blocking at 24" o.c., toe-nailed to joists

$3/4$" tongue-and-groove plywood subfloor

Floor joists, 16" o.c.

12 inches on-center, your $3/4$-inch plywood subfloor is certainly adequate, and you should have no problems. — *Howard Brickman*

Wide Plank Flooring Problems

Q. Why are my wide-board pine floors shrinking and warping?

A. There are many reasons why wide planks shrink, cup, and twist. A lot depends on how the lumber grew. But the most important factor that we can control is the moisture content of the planks. Also, the wider the board, the greater the shrinkage will be.

The floor boards will shrink if their moisture content was high at installation. After the house has been occupied for a heating season, the boards will dry and contract. In most houses in cold climates, the final moisture content of wood is 9% to 11%. If the moisture content at installation is more than 13%, significant shrinkage is likely. In the old days, flooring was always "conditioned" by storing it, stacked and stickered, in the house for 30 to 60 days before it was laid. This allowed the moisture content of the flooring to equalize with the rest of the house. Nowadays, you can measure the moisture content with a moisture meter to see if it's safe to lay the floor. If the moisture levels are high, you'll have to condition the wood.

As for warping, wet wood will warp as it dries, particularly if it is flat-sawn. Since most of the pine cut these days is from smaller, second-growth timber, the likelihood of it being flat-sawn or flat-grained is high. Edge-grain (quartersawn) boards will warp less but cost a lot more.

Moisture from below the floor, such as from a damp crawlspace or basement, can also cause the boards to cup and twist. Placing asphalt felt paper under the flooring can help, but the real cure in this case is to install a sealed vapor barrier over the damp areas. Improving the foundation drainage can also help reduce basement moisture problems. — *Henry Spies*

Felt or Rosin Paper Under Strip Flooring?

Q. I'm using a new floor finisher who is known as the best in this area. He saw that we had installed oak and ash floors over rosin paper and insisted that we should have used No. 30 felt. He says the rosin paper will degrade over time and won't provide any cushion between the subfloor and the hardwood. Is he right? We only use felt in old houses where the basement is really moist.

A. Rosin paper: bad. No. 15 and No. 30 asphalt-saturated felt: good.

Installers love red rosin paper because it is really cheap and easily covers up the dusty subfloor, making it easier to slide the wood flooring into position during installation. But those reasons don't have anything to do with the quality of the completed wood floor.

Asphalt-saturated felt, on the other hand, performs a number of functions: it retards the flow of moisture from the underside; it increases friction between the bottom of the flooring and the surface of the subfloor, resisting lateral movement during shrinking and swelling; and it provides some adhesion between the bottom of the flooring and the surface of the subfloor, helping to eliminate squeaks when the flooring is nailed properly.

And don't forget that many manufacturers and wood flooring trade associations require you to use No. 15 asphalt-saturated felt or building paper; failure to use it may be regarded as a defect if you get a complaint.

Finally, at some point new houses become old houses, and since you can't be certain about long-term moisture conditions, why not build for the long run? — *Howard Brickman*

Hardwood Strip Over Vinyl

Q. What's the right way to install hardwood strip flooring over an existing linoleum floor? The linoleum is glued to ⅝-inch plywood, which is installed over felt paper and a ½-inch plywood subfloor. The floor joists are 16 inches on-center.

A. Since a vinyl floor is not likely to create a finish floor height problem, you can go ahead and install the floor over the linoleum. If it is actually linoleum, you might want to consider pulling it up if it seems to break up when you nail into it. Be careful when lifting vinyl and linoleum — some products have an asbestos backing — another good reason to leave it in place. Also, an intact vinyl floor will provide an effective vapor barrier.

With the double layer of plywood, you should have a plenty stiff floor, and the thickness will provide good nail retention. Two-inch nails won't even protrude from the underside.

Install the hardwood floor as you normally would. I would first lay down a layer of No. 15 felt to eliminate squeaks. — *Howard Brickman*

Wood Floors Over Concrete Slabs

Q. What is the best way to install a wood floor over a concrete slab?

A. Concrete is a good substrate for installing wood flooring if proper precautions are taken to ensure that excessive moisture conditions are resolved prior to installation and controlled during the life of the floor. In addition, traditional nail-down, solid 3/4-inch, strip or plank flooring must have an adequate wood substrate for proper nailing.

Concrete substrates. Even though concrete was used to build Hoover Dam, it doesn't qualify as a "waterproof" material. Actually, the opposite is true. Concrete is quite porous.

To avoid excessive moisture problems, new slabs must be detailed properly. Place at least 6 inches of gravel or crushed stone on the ground, then install a 6-mil polyethylene vapor barrier. Make sure that this vapor barrier is not destroyed when the concrete is poured. The exterior of the concrete slab or stem wall should also be damp-proofed prior to backfilling. Before laying the floor, the slab should be allowed to dry sufficiently.

Always check for moisture in concrete prior to installing a wood floor by taping polyethylene over a clean place on the slab and allowing 12 to 24 hours for signs of moisture to develop. If the poly is clouded or beaded up with moisture, the slab is too wet.

Solid 3/4-inch tongue-and-groove strip and plank flooring cannot be installed directly to concrete. You must install a wood subfloor for nail-

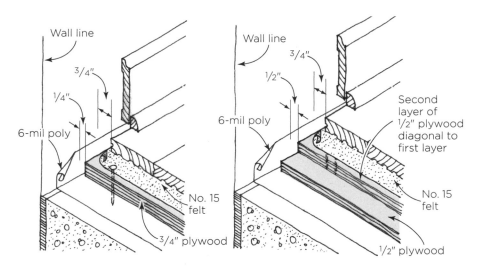

Wood floors over concrete: Solid-wood strip or plank flooring must have an adequate wood substrate for secure nailing over concrete. Use either the "plywood-on-slab" method (left) or the "floating plywood" system (right).

ing. Two recommended methods are a 3/4-inch "plywood-on-slab" and a "floating plywood" system.

Plywood-on-slab. Place a 6-mil polyethylene vapor retarder directly over the concrete. Lay 3/4-inch plywood sheets opposite to the direction of the finish flooring, leaving a 1/4-inch gap between panels to eliminate squeaking. Fasten the plywood to the concrete using powder-actuated fasteners or concrete nails. Standard 2-inch power-cleats or pneumatic staples will contact the concrete surface beneath the plywood unless the machines are tilted forward by placing a 5/16-inch-thick spacer at the back edge of the nailer's faceplate. The alternative is to purchase 1 3/4-inch power-cleats specifically designed for this application.

Floating plywood system. This system is recommended where there is any question about moisture coming up through the slab. Place a 6-mil polyethylene vapor barrier over the concrete. Lay 1/2-inch plywood sheets along the long axis of the room. Place a second layer of 1/2-inch plywood at 45 degrees over the first layer. Again, leave 1/4-inch gaps between sheets of plywood to eliminate squeaks. Then fasten the two layers of plywood together using 7/8-inch pneumatic staples. Leave a minimum of 1/2-inch expansion space at the perimeter for normal room sizes. It is a good idea to increase expansion for large areas. If the concrete slab is uninsulated, you can install a layer of 1-inch-thick foam insulation between the slab and the plywood, with the 6-mil poly over the foam. Use tongue-and-groove extruded polystyrene with a high compressive strength (rated for use under slabs or radiant floors).

Laying the floor. When laying any wood floor, be sure to:
- Use No. 15 asphalt-saturated building paper or felt under all nail-down floors
- Start the flooring straight using a chalk line or string
- Use plenty of nails
- Make use of the tongue and groove or splines when changing direction or working from one room to another.

— Howard Brickman

Termites Love Oak Floors

Q. We are repairing an oak strip floor. Several boards obviously need to be replaced because of visible termite damage. Is there some way to evaluate the unseen damage to the rest of the floor?

A. Oak seems to be one of termites' favorite foods, and they will go to great lengths to reach it. About all you can do is probe the wood with a sharp object. I use a small screwdriver, sticking it into the summer wood (the light-colored grain). If termites have infested the floor boards, they will have eaten the summer wood and not the dense, darker grain. Termites generally follow the length of the board rather than working across the grain. If the board will support the screwdriver blade without crushing or splitting, it should be strong enough to survive without replacement. Even spike heels will not equal the pressure of the screwdriver blade. — *Henry Spies*

Finishing a Pine Floor

Q. I am installing 1x12-inch-wide eastern white pine flooring throughout a new home, including the kitchen. Since the flooring is used in high-traffic areas, I want to choose a finish that will provide long service. What do you recommend?

A. If the customer expects a finish that will provide long service, they are likely to be disappointed. In this case, your most important job is to lower the expectations of your client.

Eastern white pine *(Pinus strobus)* is a low-density species. It has a specific gravity of 0.35, which is roughly half as dense as oak or maple. You need to prepare your customer for the eventual denting and scratching that is characteristic of pine floors.

Pine floors get scratched when they are walked on. There is no such thing as a no-maintenance wood floor, so the floors will need a fresh coat of finish when traffic marks start to show. Choose a low-sheen finish, since the higher the gloss, the more quickly scratches and scuffs will become visible. — *Howard Brickman*

Water-Based Polyurethanes vs. Oil-Based

Q. Please comment on the durability of water-based polyurethanes for wood floors. Will these products hold up in heavy traffic areas as well as the solvent-based products?

A. I am confident that water-based finishes will perform as well as or better than oil-based products if the installer takes the time to master some new application techniques.

Waterborne finishes have been used for years in high-traffic areas, such as gyms and racquetball courts. Waterborne floor finishes have a low viscosity — that is, they're watery — so the finish penetrates deep into the wood. While this increases their durability, it can also create some application problems that you've got to master.

For starters, the finish can really raise the grain on open-grained wood, such as oak. To combat this problem, I mist the floor very lightly with a two-gallon garden sprayer, not allowing the water to pool. Then I let the wood dry for about 12 hours (check with a moisture meter to be sure the moisture content has not been raised). This is sufficient to raise the grain, which I knock down with a worn screen back. If you do this before applying the first coat, you'll avoid problems caused by excessive disking between coats.

Because you have less buildup, disk between coats with a light abrasive pad, instead of a screen back. These pads are color-coded, and green seems to work best.

Also, coats of waterborne polyurethane are thin and dry quickly, so it's important to keep a wet edge when laying them down. To make this happen, you may need to add a retarder (ask the coating manufacturer), lower the heat, and keep doors and windows closed to reduce air movement while applying the finish. Then allow the finish to dry as slowly as possible to cure it.

Some finishes are marketed as quick-dry products, but ignore this marketing hype. The better the coat is cured, the stronger it will be.

— *Michael Purser*

Recoating Wood Floors

Q. We're renovating a home with hardwood floors that are somewhat worn and lackluster. Is there a way to restore the finish without sanding?

A. I recommend that you consider recoating the surface. It offers great value and a quick turnaround and should be completely dustless. You can recoat any contemporary finish as long as paste or acrylic wax has not been applied over it. I further recommend that a professional floor finisher do the work because professionals have the appropriate equipment and experience.

After the rooms have been cleared of furniture, prepare the surface with a buffer and a maroon synthetic pad or well-worn 120 screen-back disc. Simple vacuum attachments are available for the buffer to collect all the dust created. You might also consider using a chemical prep, which eliminates buffing entirely. Either way, there need not be any dust.

After preparing the surface, you apply the finish. I suggest waterborne polyurethane because of its relatively fast drying and curing time, its ease of maintenance, and its attractiveness. If you have any high-wear areas, you might want to touch them up before making the final application. After that, one coat usually does the job.

Understand that a recoat does not remove serious gouging or indentations, nor will the preparation remove paint or stains. But, it provides excellent long-term protection and enhances the appearance of the floors with a lot less effort and money than a full-blown sanding and refinish. — *Michael Purser*

Refinishing Faded Wood Floors

Q. My 7-year-old hardwood floor needs to be refinished because the stain faded from the sun. I am hoping to restain it with a product that can withstand direct sunlight. Do you have any recommendations?

A. Floor finishes with ultraviolet (UV) inhibitors and stains with pigments (as opposed to dyes) will slow down color degradation in both the finish itself and the underlying wood. And UV light isn't the only cause of fading and discoloration; heat and moisture can also be factors. To slow down (notice I didn't say "prevent") fading and deterioration, you'll have to manage all three causes.

The key is to avoid extremes of exposure to sunlight and humidity. When building a new house or adding on, your clients would be wise to consider low-E glass. Made of transparent metallic oxides, low-E coatings can reflect up to 90% of long-wave infrared (IR) light, the kind that creates the heat that accelerates oxidation in wood. But typical low-E glass only stops about 50% of ultraviolet light, the type most responsible for fading. A good choice is so-called "spectrally selective" low-E glass, which blocks nearly 70% of the light frequencies that cause fading. Special UV coatings and tinted glass can provide better performance but are used primarily in commercial projects.

If you have to work with older existing windows, you might want to investigate window films. Often applied to the interior glass of historic properties, these thin, multilayered film products have various coatings that can selectively reduce UV, IR, and visible light, all of which contribute to fading.

Finally, there's a low-tech solution: window coverings. Drapes, shutters, and blinds were put to good use by previous generations. Compared with refinishing your floor or retrofitting new windows, window coverings are also fairly cheap. The bottom line is that sunlight is destructive to wood as well as to the various finishes used to protect it; you can slow fading down, but you can't stop it. So your focus should be on the light entering the room and how to reduce its intensity.
— *Michael Purser*

OSB as Carpet Substrate?

Q. Why can't I use $7/16$-inch OSB over 1-inch rough pine boards for carpet underlayment?

A. There's no reason you shouldn't use OSB, as long as there's no risk of it getting wet (if it does, it swells and comes apart). First, make sure that the diagonal board subflooring is screwed down and that there are no squeaks. Then lay the OSB in plenty of construction adhesive and screw it to the subflooring. The carpet should install just fine.   — *Ken Smith*

Carpet Over Ceramic Tile

Q. We are converting a restaurant kitchen into an office and need to install carpet over existing 6x6-inch ceramic tiles with $1/4$-inch grout lines. The floor dips $3/4$ inch near the floor drain. What product can we use to level this floor in preparation for a pad and carpet?

A. Before proceeding, verify that the moisture level of the substrate is not excessive. Remove any loose tiles or grout, and cap the drain. Clean the tiles to remove any grease, wax, or coatings that might interfere with adhesion. Then install a Portland-cement-based self-leveling underlayment, following the manufacturer's recommendations. Self-leveling underlayments are available from Ardex (724/857-6400, www.ardex.com) and Mapei (888/300-4422, www.mapei.com).

Once the floor is leveled, cured, and sealed, the carpet can be installed. If you want to include a pad, you have two options: either use "double-glue cushion" — a type of pad that can be glued to both the substrate and the carpet — or install carpet with an attached cushion backing. Do not nail tack-strip into the hardened leveling compound unless you are sure it is at least $3/4$ inch thick, since nails can shatter the compound where it is thin.   — *Patricia Davidson*

Vinyl Flooring on Slabs

Q. When installing vinyl flooring on a below-grade slab, how do I check for excessive moisture in the slab?

A. There are two "subjective" tests that can be used to check for the presence of excessive moisture levels in a slab on grade.

The simpler is the "Rubber Sheet/Plastic Mat Test." Using duct tape, carefully tape the edges of a 2x2-foot piece of 4-mil poly to the slab for 72 hours. If you observe any moisture droplets or surface color darkening of the slab, excessive moisture is present.

A more definitive approach is the "Bond and Moisture Test." Install a 3x3-foot piece of the actual flooring material per specs, using the proper adhesive. Seal the perimeter of the test area tightly to the floor with duct tape and allow it to cure for at least 72 hours. Remove the tape, and pry the flooring from the slab. You have a moisture problem if the tile comes up easily, the adhesive appears damp and stringy, or the adhesive releases from the concrete and stays on the flooring material.

There is also a scientific test, called the "Calcium Chloride Test," which measures moisture migration rates using a petri dish covered by a plastic dome. If this test is to be used, contact the flooring manufacturer for the range of acceptable values. — *Jim Hamrick*

Solid Surfacing for Flooring

Q. Can solid surfacing be used for flooring in a bathroom?

A. Some solid-surfacing fabricators have used solid surfacing for flooring despite the fact that most manufacturers, including DuPont, the manufacturer of Corian, will not provide a warranty for flooring applications.

Besides the high cost, one disadvantage of using solid surfacing for flooring is its surface texture. Solid surfacing is so smooth that it's slippery when wet. But if the material is cut into 12x12-inch tiles and installed with grout lines, it will be somewhat less slippery than a seamless application. — *Martin Holladay*

Floor Tiles vs. Wall Tiles: What's the Difference?

Q. One of the tile showrooms where my remodeling clients shop has tiles displayed in two sections, one for floors and one for walls. What's the difference between a floor tile and a wall tile? Is it possible to interchange the two?

A. On the surface, a tile's suitability for a particular application may not be readily apparent. The strength and durability of a tile are determined by a few factors: the ingredients making up the bisque, or body, of the tile; the type of glaze used, if any; and how long and at what temperature the tile is fired. Wall tiles often have decorative or high-gloss glaze applications that are really appropriate only on wall surfaces. Floor tiles, on the other hand, may be used for either floor or wall applications.

How to tell the difference? The best way is to ask the person selling the tile for a written recommendation. Most of the time, this comes in the form of a manufacturer's brochure that states that a certain tile is approved by the manufacturer for use on floors and walls, walls only, or floors only. There may also be a statement regarding a tile's use on a countertop. If no such written recommendation is available, a written statement on the tile seller's company letterhead will suffice.

Keep in mind that some porcelain floor tiles have a somewhat granular surface meant to reduce slipping. Some of these tiles are also manufactured specifically for use in commercial applications where heavy-duty cleaning machinery will be used to maintain the floors. Because such heavy cleaning is impractical in a home or office, the textured surface may get dirtier than if it were installed in a commercial or industrial setting.

While it is important to know the wear qualities, a tile's absorbency should also be considered. Tiles that are bulletproof in one application may not be appropriate for another. For example, Saltillo and other hand-molded paver tiles are used extensively for both indoor and outdoor floors (in nonfreezing climates), but they should not be used in wet interior applications where hygiene is an issue because they are generally too absorbent.

Some tile sellers may refer to the hardness scale for assessing the appropriateness of a tile for floor use. While that might be helpful, with today's new glazes it may not be as reliable as it once was. Hardness is important, but surface texture may be a better indicator. For example, some porcelain tiles that are extremely hard would be too abrasive for normal home or office use.

Don't rely on anecdotal "advice" if you don't have specific experience with the tile you choose: ask your dealer and get an assurance in writing if you have any doubts. — *Michael Byrne*

Subfloor Under Tile

Q. We will be installing ceramic tile flooring in the kitchen of an existing house. The subfloor consists of 1/2-inch plywood over 3/4-inch particleboard (not OSB). The joists are spaced 16 inches on-center. My plan is to install 1/2-inch cementitious backerboard on the plywood and then install the tile. Will this be adequate?

A. Particleboard, which easily absorbs moisture and is dimensionally unstable, shouldn't be included anywhere as part of a tile substrate. In this case, the existing plywood and particleboard subfloor layers will have to be removed and a new subfloor installed on the joists.

The best subfloor for a backerboard and tile installation is 3/4-inch plywood, glued with subfloor adhesive and mechanically fastened every 8 to 12 inches in the field and every 6 inches on the edges. Leave a 1/8-inch gap between plywood sheets for expansion relief. Then install your 1/2-inch backerboard in a freshly combed bed of thinset (to give uniform support), fastening the backerboard with corrosion-resistant fasteners. Since installation requirements vary depending on the backerboard manufacturer, always verify a manufacturer's requirements before proceeding. — *Chip O'Rear*

Air-Nailing Backerboard

Q. Can I use a pneumatic nail gun to fasten tile backerboard to floors and walls?

A. ANSI (American National Standards Institute) specifications require that a corrosion-resistant roofing nail be used when nailing tile backer-board, and that the nail penetrate the wood framing at least 3/4 inch. While there are no direct references to pneumatic nailing, I see no reason why pneumatic roofing nails couldn't be used, as long as the fasteners were of sufficient length. It's important that the nail head does not break the fiberglass matting material embedded in the backerboard, so make sure to properly set both the gun's depth adjustment and the air pressure.

Corrosion-resistant screws can also be used, providing their head diameter is large enough to resist a 125-pound pull-through force. ITW Buildex's Hi-Lo/Rock-On screws meet this requirement (800/323-0720, www.buildex.com).

Backerboard installed over a plywood subfloor should always be fully bedded in mortar or adhesive. In these situations, many contractors mistakenly think that the fasteners serve as clamps and that their holding power is not an issue after the mortar or adhesive "grabs." But the primary function of the setting bed is to provide a leveling bed for the backerboard; the fasteners are what holds the backerboard in place for the life of the tile installation.

For detailed guidelines on tile installation requirements, contact the Tile Council of America (864/646-8453, www.tileusa.com) for a copy of the *Handbook for Ceramic Tile Installation*.   — *Michael Byrne*

Ceramic Tile on Basement Slab

Q. What's the best way to vapor-seal a basement slab before tiling over it?

A. Surface-sealing a slab before applying tile is not required or recommended by the tile industry.

Tile can readily tolerate basement slab moisture, as long as you choose the appropriate tile and tile-setting material. In fact, applying sealer to concrete is likely to cause problems for tile by closing the pores that are required for cement-based compounds to bond to the concrete slab. In any case, you definitely cannot use mastic, premixed thinset, or any latex-modified product in this application.

If the slab is perfectly sealed, the material will not adhere because the pores will be closed; on the other hand, if any moisture does rise out of the slab, the adhesive will probably deteriorate and lose the bond that way.

For ceramic tile in a basement with a history of moisture issues, your best choice would be a vitreous tile (with a water absorption of .5% to 3%) or a semivitreous tile (with a water absorption of 3% to 5%). That would allow the use of regular dryset mortar instead of the latex- or polymer-modified material recommended by manufacturers of porcelain tile, which is impervious (less than .5% absorption). Standard dryset, unlike many latex and polymer formulations, will cure very well in a damp environment. Make sure the surface is clean, free of sealers, and free of any standing water before application.

One caution: In most instances of high moisture, alkalinity is also present, which may cause efflorescence (a white powdery deposit that typically appears first at tile edges). If the slab is alkaline, talk to your local concrete and masonry supply house about treating the slab with a cleaning product like Sure Klean Restoration Cleaner from Prosoco (800/255-4255, www.prosoco.com). But if the alkalinity is continually migrating through the slab, it may be caused by excessive subsoil moisture. In such instances, chemical treatment will not have a lasting effect and you may have to take more expensive measures, such as drainage improvement or, in rare cases, slab reconstruction.   — *Dave Gobis*

Reinforcing Mud-Bed Tile Floors

Q. What's the right type of mesh to use in a mud-bed tile floor? Should I use a self-furring lath (like the type used in stucco work) so that it centers itself in the mud?

A. First, there are several mortar bed installation methods that require no reinforcing. These are typically for floors where you're putting a bonded mortar bed over an intact slab-on-grade. These methods, which are based on ANSI A108 specification standards, are described and illustrated in the *Handbook for Ceramic Tile Installation*, available from the Tile Council of America (864/646-8453, www.tileusa.com).

I'll assume you're talking about laying a mortar-bed tile floor over wood framing. The tile industry recognizes two methods for reinforcing mortar setting beds for floor tiles over wood-framed floors (F141 and F145 in the TCA handbook). The difference in the two methods has to do with the thickness of the mortar.

For thick bed installations — from 1 1/4 inches to 2 inches thick — the ANSI A108 specification, A-2.1.7, calls for one of the following welded wire fabrics:
- 2x2-inch x 16/16 wire
- 3x3-inch x 13/13 wire
- 1 1/2x2-inch x 16/13 wire
- 2x4-inch x 16/16 wire

For this type of installation, the thickness of the mortar requires that the reinforcing fabric be positioned somewhere in the middle of the mortar bed. When I set tile using this method, I'll dump about half the mortar on the floor, lay the wire mesh on top of that, then spread the rest of the mortar. This works fairly well with mud-bed mortar because it's fairly dry and will support the wire. It isn't necessary to use any other supports for the wire.

The thin mortar bed method (3/4-inch-minimum thickness) calls for flat, expanded metal mesh weighing not less than 2.5 pounds per square yard. Painted lath is allowed, but galvanized lath is preferred. The TCA F-145 detail shows the lath fastened snugly to the subfloor, with a cleavage membrane installed between the two.

In either the thin or thick bed method, a cleavage membrane, which

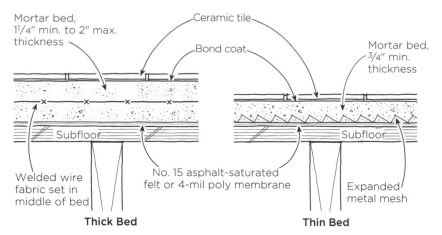

Mortar bed, 1¼" min. to 2" max. thickness

Ceramic tile

Bond coat

Mortar bed, ¾" min. thickness

Subfloor

Subfloor

Welded wire fabric set in middle of bed

No. 15 asphalt-saturated felt or 4-mil poly membrane

Expanded metal mesh

Thick Bed

Thin Bed

For thick mortar bed installations over wood floors (at left), ANSI specs call for wire fabric reinforcing. For thin mortar beds (at right), use expanded metal lath nailed or stapled to the subfloor. Both methods require a cleavage membrane between the subfloor and the mortar.

can be No. 15 asphalt-saturated roofing felt or 4-mil poly, separates the mud from the wood floor and prevents the wood subflooring from drawing moisture out of the mix during the curing phase.

I prefer mortar beds for tile floor installation because they provide a smooth, flat surface for tile setting, but they require specific detailing beyond the type of mesh to use. If you plan to use mortar-setting beds, get a copy of the TCA Handbook for a review of all the details. — *Michael Byrne*

Tile Over Painted Concrete

Q. Can I install ceramic tile over a painted concrete floor without removing the paint?

A. Unless the paint in question is properly applied epoxy paint (not the kind that comes in a spray can), you shouldn't install ceramic tiles over either paint or paint overspray. If tiles are installed over a painted floor, the bond strength of the tile adhesive will not be any stronger than the paint's grip on the concrete. — *Michael Byrne*

Tiling Over Old Linoleum

Q. Is it okay to linoleum over an old linoleum tile floor? I'm concerned the old linoleum might contain asbestos.

A. There may be advantages to tiling over some existing flooring materials that contain asbestos, provided that the existing structure and its subflooring can be identified and confirmed as adequate for the additional weight of a tile installation.

More and more communities are giving the okay to covering existing asbestos-bearing flooring materials with ceramic or stone tiles, as long as the structure is up to the task of bearing the weight. If so, the common practice is to install 1/4-inch-thick tile backerboard over the asbestos material, perhaps add a waterproofing/crack-isolation membrane over the backerboard, and then install the tiles with latex thinset mortar and a grid of movement joints filled with a resilient sealant instead of grout.

The movement joints should extend from the top surface of the tile to the bottom of the underlayment, but they should not extend down into, or penetrate, the suspect material. The width, number, and placement of movement joints (which are necessary in every tile installation, without exception) are too complicated for a brief explanation, but the TCA *Handbook for Ceramic Tile Installation* contains a thorough discussion under Detail EF171 (Tile Council of America, 864/646-8453, www.tileusa.com).
— *Michael Byrne*

Cleaning Grout

Q. I have a client who had a tile floor put down a few years ago. The tile installer told her that the grout already had a sealer in it so no additional sealer was ever put on. Now the grout is stained, and she has been trying to clean it with bleach (the grout is white) but cannot stand the fumes. Is there anything that can be used to clean the grout that would be fairly simple to use and not smell too bad?

A. I have never used or heard of a grout that contains a built-in sealer. Although latex and epoxy grouts are somewhat more stain-resistant than regular Portland cement grouts, a quality high-performance sealer should always be applied — after the grout has cured — to help reduce staining and to make housekeeping easier.

White grout is a poor choice for any tile installation intended for a heavy-use or a food service area, and without a sealer, it may turn "antique" white or even gray. Bleach can sometimes be used to remove stains from white grout, but it will fade colored grout.

Once grout has been cleaned, it should be allowed to dry thoroughly and then protected with an appropriate sealer.

The first step in the cleaning process is to identify the grout. Many grout manufacturers offer lines of cleaning and sealing products that, when used as a single-source system, can significantly reduce fading and color change. If an after-market sealer is used, I choose a manufacturer that produces both a sealer and a cleaning preparation. Some acid-based cleaners may require one or more wash-rinse neutralizing cycles before a sealer can be applied.

On new installations, I never use regular grout, which is a mixture of Portland cement, sand, and water. Instead, I replace the water with a liquid latex or acrylic, and mix it with regular grout powders (sanded or unsanded). In some cases, the latex component is factory-added as a dried powder — called polymer-modified grout — and water is the liquid component.

After the tiles are cleaned, and once the fresh grout joints have begun to set up, I strike the joints to produce a smoother grout surface. When hard, cured, and protected with an impregnating sealer, smooth-surface joints are significantly easier to clean than joints whose surfaces are rough or uneven. Striking fresh grout is one of the secrets of easy maintenance for tile. Adding the protection of a quality sealer makes this kind of finish ideal for an installation whose appearance is important.   — *Michael Byrne*

Removing Tiles From Plywood

Q. What is the best way to remove mastic-adhered ceramic tiles from plywood without ruining the subfloor?

A. I would try an electric floor stripper with a tile blade to remove the tile, then a spring steel blade to scrape off as much of the mastic as possible. An electric floor stripper is a rather specialized tool, but it can usually be rented from a tool rental company or a floor covering supplier.
— *Henry Spies*

Steam Room Details

Q. My customer wants to have her combination steam room/shower tiled with a nonglazed floor and wall tile she has already purchased. Is the tile suitable for this type of application? I would use backerboard for a substrate and the appropriate thinset mortar.

A. This is one of the most difficult of all tile installations to properly specify and install. First, I would stay away from using any unglazed tile

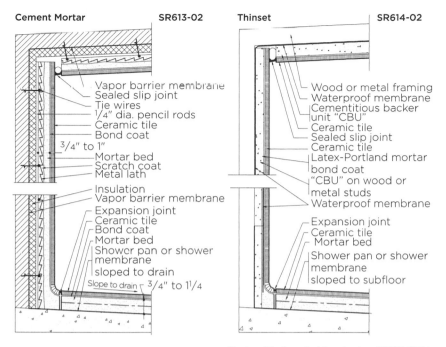

| Cement Mortar | SR613-02 | Thinset | SR614-02 |

Vapor barrier membrane
Sealed slip joint
Tie wires
1/4" dia. pencil rods
Ceramic tile
Bond coat
3/4" to 1"
Mortar bed
Scratch coat
Metal lath

Insulation
Vapor barrier membrane
Expansion joint
Ceramic tile
Bond coat
Mortar bed
Shower pan or shower
membrane
sloped to drain
Slope to drain 3/4" to 1 1/4

Wood or metal framing
Waterproof membrane
Cementitious backer
unit "CBU"
Ceramic tile
Sealed slip joint
Ceramic tile
Latex-Portland mortar
bond coat
"CBU" on wood or
metal studs
Waterproof membrane

Expansion joint
Ceramic tile
Mortar bed
Shower pan or shower
membrane
sloped to subfloor

Courtesy Tile Council of America, Inc., ©2003–2004

in this area except for the floor, where slip resistance is desired. Unglazed tiles in a steam room may cause maintenance problems because salts tend to build up in such areas; salts are more easily cleaned off glazed tiles and may prove difficult to remove from the surface of unglazed tiles.

Have you checked with the backerboard manufacturer to see if its product can be used in a steam room environment? Some backerboards may not be suitable for such use.

You say nothing about how the floor of the shower will be built. Certainly, you should not expect to use backerboard there, since the floor must be sloped to the drain. Also, cement backerboards are known to wick moisture: Therefore, the junction between the wall and the floor setting bed will require a waterproofing detail that may differ from one brand of backerboard to the next.

On steam units, I always recommend a surface-applied waterproofing membrane (liquid or sheet). Membranes located behind the backerboard will allow steam vapor to enter the wall cavity.

I suggest that you obtain a copy of the TCA *Handbook for Ceramic Tile Installation* (864/646-8453, www.tileusa.com) for details on steam showers and other ceramic tile installations. In the meantime, see the illustration, which provides some basic specifications to follow.

— *Michael Byrne*

Sealing a Shower Drain

Q. When installing a mortar-bed shower floor, should the gap between the drain and the adjacent tiles be finished with grout or with caulk?

A. Use grout. Since the shower floor, if installed properly, has a mud base (Portland cement and sand) at least 2 inches thick, there should be almost no potential for deflection. Caulk is used mostly at the joint between vertical and horizontal surfaces, where movement is more likely. One reason to limit the use of caulk is that many caulks will mildew and discolor in wet areas.   — *Tom Meehan*

Cracking Tiles in Shower

Q. I have been called to inspect a shower with ceramic tile walls. The substrate is cementitious backerboard, and as far as I can tell, the tile was installed with thinset. There are hairline cracks running through the tiles on all three walls, both vertical and horizontal. There is a ceramic soap dish in the corner, and even it is cracked. It doesn't appear that there has been any movement in the floor or the walls. What could be causing the cracking?

A. Tiles generally crack for only two reasons: loss of bond or moving substrate. In this case, the substrate might be moving if the studs are spaced farther apart than 16 inches on-center, allowing that backerboard to flex.

Another possibility is that the backerboard was cracked prior to installation. Although cementitious backerboard often has minor cracks that don't cause problems, over-stressing the boards during delivery or installation can cause more serious cracks. If the backerboard was installed over a layer of drywall, then water penetration through tile joints or corners may have soaked and softened the drywall.

If the tiles have parted from the setting bed, they can break when someone applies pressure on the poorly supported tile. A tile can also crack if it is installed over two different substrate materials or an expansion joint. Tiles are strong only when they are cushioned in a bed of adhesive and surrounded by grout.

Solving the problem will require removing the tiles so that the substrate can be inspected.   — *Michael Byrne*

Countertops

Substrates for Tile Countertops

Q. Can countertop tiles be installed directly to a plywood base?

A. If the tiles in question will be used to cover a serving counter that is normally dry, then they can be set directly over plywood with an epoxy mortar or an organic adhesive. But if a sink is part of the installation, plywood should not be used as the base for tiles.

The American National Standards Institute (ANSI) specification for tile characterizes plywood as dimensionally unstable and not ideal backing for ceramic or stone tiles. Nevertheless, many thinset mortar manufacturers produce tile adhesives specifically for use with plywood.

In my work, if plywood is specified in plans and must be used as the base for tiles, I get the architect or GC to include a waterproofing membranc. I recommend either a sheet membrane, such as the Noble Company's NobleSeal TS (800/878-5788, www.noblecompany.com), or a trowel-applied membrane, such as Laticrete 9235 (Laticrete International, 800/243-4788; www.laticrete.com).

A better approach than using plywood, however, is to use cement backerboard, which is made specifically for tiles. Cement backerboards provide a tough base for ceramic or stone tiles, and are unaffected by water or moisture.

For moderate protection against moisture penetration, the backerboard can be laminated directly to a plywood base with latex-modified thinset mortar. For a commercial or active residential kitchen or bathroom countertop, however, I recommend combining the backerboard with a waterproofing membrane as the base for the countertop tiles.

— *Michael Byrne*

Preventing Grout Stains on Kitchen Counters

Q. I often hear clients complain that they like the look of their tile kitchen counters, but that it's too hard to keep the grout lines clean. Is there any kind of grout that resists staining, or a sealer that works?

A. The problem with discolored grout in kitchens is all too common, but it doesn't have to be that way. There are three main issues: Choosing the right grout and grout color, installing it correctly, and sealing it with the right product. Tile choice can also make a difference.

Before choosing a grout color, you need to know how heavily the kitchen will be used. In a kitchen that sees only light cooking and cleanup, you should be able to use any color. Use pink if it makes the client happy. But if a lot of cooking is planned, you should probably choose a grout the color of dirty dishwater — gray — because that's the color the grout will want to turn. If you put a white grout in that kitchen (the worst choice), you're essentially making the homeowner a slave to keeping it clean. Black and other very dark colors are also bad because they show up light stains and deposits. Steer your client toward cement gray. The most durable grout is an epoxy, which is available in gray. But if your client just has to have that dark green grout she's seen in a friend's kitchen, use a latex grout — you'll have a better color choice.

Next, you have to apply the grout properly. Make sure you mix it according to instructions, and make sure to use a grout trowel so the material is forced into and completely fills the joint.

A step that is usually neglected is striking the joint. Most installers use only a sponge for grouting, which has the effect of removing the fine Portland cement and sand filler and exposing the sharp edges of the silica aggregate — leaving a very rough and abrasive grout surface. Imagine dropping a peanut butter sandwich on a piece of sandpaper and trying to clean it — this is what the homeowner will be faced with in trying to clean that grout surface. Striking the joint with a tool like a brick mason uses will compact and smooth the grout, making for a much smoother, less absorbent, and easier-to-clean grout line.

If you don't install the grout properly, sealing is a useless exercise. But assuming you've tooled the grout correctly, now apply an impregnating sealer. A good choice is 511 Porous Plus (Miracle Sealants, 800/350-1901; www.miraclesealants.com). This is a solvent-based sealer used on stone buildings to make it easy to clean off graffiti. It requires 72 hours of curing, but once cured, it's food-safe. Two other good choices would be Sealer's Choice 15 or Grout Sealer from AquaMix (877/278-2311, www.aquamix.com). Grout Sealer is an acrylic product that leaves a smooth, easy-to-clean surface; Sealer's Choice is a commercial-duty product with a 15-year service life. I would avoid using silicone sealers, which are ineffective.

Sealers have to be maintained, depending on use. And tile needs to be cleaned for staining just like any other surface — if you spill some wine in a grout line, sponge it up. Tile choice is not usually the contractor's decision, but if your client wants a tile counter and is concerned about grout staining, you might recommend using oversized tiles on the counter to reduce the number of grout lines.   — *Michael Byrne*

Removing Old Grout

Q. Is there an easy way to remove old grout without working by hand with a grout saw?

A. If the reason you're removing the grout is that it's old and crumbly, try a utility knife first. If the grout is soft enough, this will work, though you may go through a lot of blades. For wide grout lines — 1/4 inch or more — you might try this technique: with a steady hand, use a drycutting diamond blade in an angle grinder or circular saw and cut out the middle of the joint. Then cut away the rest with a utility knife (set the blade only as deep as the tile).

For really persistent grout and narrow joints, there's no easy answer: it's a tedious and time-consuming process that's unforgiving of mistakes, such as slipping and scoring the tile.

If the reason you want to remove the grout is that it's discolored, you might spare yourself the effort and try a grout colorant to hide the stain instead.
— *Michael Byrne*

Stripping Laminate

Q. I need to remove old Formica in a house we're restoring. I've had limited success using heat guns for this task. Any recommendations?

A. I've had good luck using acetone to remove glued-on laminates. I fill a spray bottle (use glass — a plastic bottle will melt) with acetone and spray a mist along one edge of the laminate. After the acetone has soaked in for about a minute, I start to work the laminate loose with a putty knife. I continue to spray acetone into the widening gap, repeating the process until the entire piece comes off.

Using this method, a helper and I have removed 30 square feet of laminate from a kitchen countertop in about 40 minutes.   — *Bruce A. Wooster*

Rx for Delaminating Countertop

Q. What type of glue is best for repairing laminate that is beginning to lift at the edge of a kitchen countertop?

A. If the existing glue is contact cement (which is typical), then I would recommend using that. Just apply a generous amount of contact cement to both surfaces and allow it to dry, then press back into place. If wood glue was used, spread water-based wood glue in the lifted area and add pressure for a couple of hours or until the glue is dry.   — *Mark A. Roberts*

Laminating Over Plastic Laminate

Q. Is laminating over existing plastic laminate an acceptable practice? If so, what is the best procedure?

A. It is possible to add a second layer of laminate, though it's always best if you can remove the existing laminate and get down to a clean substrate. Don't ever apply a second layer over loose laminate.

If the first layer is well adhered, scuff the surface with a belt sander and an 80- or 100-grit belt. The key is to break the slick surface, since most glues will not stick well to slick, nonporous surfaces. After scuffing the surface, wipe it with a solvent like mineral spirits or paint thinner. This helps the glue to "wet out" — meaning that the glue will spread well instead of beading up on the surface. After wiping the counter clean and dry, apply the laminate as you usually would.   — *Jeff Pitcher*

Adhesive Caulk for Backsplashes

Q. I've heard that an adhesive caulk does not bond as strongly as glue. Is an adhesive caulk like Phenoseal strong enough to attach a kitchen backsplash to drywall?

A. Normally, an adhesive caulk wouldn't be as strong as glue, but they are really intended for different purposes. Glue should be used in cases where two surfaces can be brought into intimate contact and held there under pressure for an extended amount of time. Often, this just isn't possible on the job site. Thus, the reason for adhesive caulks and mastic adhesives: they have a gap-filling ability not available with traditional yellow or white glue.

Adhesive caulk should certainly be strong enough to adhere a backsplash to drywall. The limiting factor is really the strength of the drywall.
                                                               — *Jeff Pitcher*

Granite Tile Countertops

Q. We are planning to use 1-foot-square granite tile in a kitchen counter application because it is more affordable than a granite slab. Can I install the granite

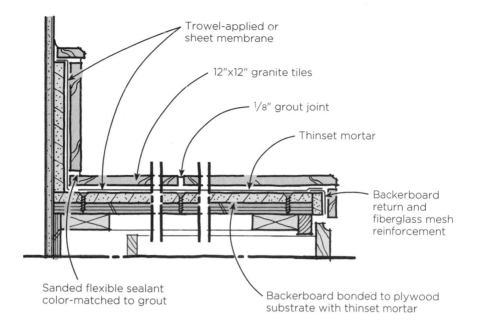

Trowel-applied or sheet membrane

12"x12" granite tiles

1/8" grout joint

Thinset mortar

Backerboard return and fiberglass mesh reinforcement

Sanded flexible sealant color-matched to grout

Backerboard bonded to plywood substrate with thinset mortar

right over plywood, or do you recommend Durock as an underlayment? What type of grout would work best? How wide a space should we leave between tiles?

A. I wouldn't risk granite tiles over plywood. The best base to use is a backerboard made specifically for ceramic tile. This could be a cementitious backer unit, like Durock or Hardibacker, or a glass-mat water-resistant gypsum board, like DensShield.

The backerboard should be laid over a 3/4-inch exterior-grade plywood substrate, attached with a leveling bed of thinset mortar in between, and secured with Hi-Lo/Rock-On screws or galvanized roofing nails. Avoid the use of drywall screws.

For a counter that will get heavy use, you should also consider using a trowel-applied or sheet membrane over the cementitious backerboard and extending up into the backsplash area (DensShield needs no further waterproofing).

As for grout, my recommendation is a natural cement-colored grout mixed with a latex additive (or use a polymer-modified grout). Width is up to you. I usually recommend narrow joints (1/8 inch or less) because they're easier to clean than wide joints.

Use a heavy-duty sealer as well (not a cheap silicone pour-on), and most important, include caulked expansion joints around sinks, cooktops, chopping blocks and at the joint between the top and the backsplash (see illustration).

— *Michael Byrne*

Sealing Stone Countertops

Q. How should I seal marble and granite countertops? To be used in a kitchen, a surface should be waterproof, heat resistant, stain resistant, and safe for food preparation.

A. All types of stone can be sealed to prevent absorption of liquids. We use VIC Lithofin PSI Premium Silicone Impregnator (available from Leonard Brush and Chemical, 502/585-2381; www.leonardbrushand chemical.com). Another good sealer is 511 Porous Plus (Miracle Sealants Co., 800/350-1901; www.miraclesealants.com), which works well with all porous stones. Yearly sealing is recommended, along with a monthly maintenance polish. These sealers are all safe for food contact once they have dried.
— *Anita Aikey-Socinsky*

Securing Dishwashers to Stone Counters

Q. I've been unsuccessful in finding a permanent method of securing dishwashers to the underside of stone countertops. I've tried using construction adhesive and epoxy to secure the small metal tabs, but with no long-term success. One countertop installer said he uses a hammer drill to drill a hole about half the thickness of the countertop. He inserts a cut-off plastic anchor in the hole and secures the tab on the dishwasher with a screw. This seems extremely risky. Do you have any suggestions?

A. I do it the same way as that countertop installer, and have never had any problem. The metal tabs on top of the dishwashers are usually far enough back that there's no risk of creating a stress crack at the edge of the counter.

I nearly always install $1^1/4$-inch-thick counters. On thinner, $3/4$-inch counters, we double the front edge thickness and carry it far enough back that we can safely drill into that. If you're worried, you could try using a bead of silicone instead of drilling. Wedge the tab into place over the silicone until it sets up. Some installers use epoxy anchors, which require a bigger hole. However, the installed anchor protrudes down about $3/8$ inch, which can get in the way of the top-mount controls on a lot of the new dishwashers.

Another way to avoid drilling the stone is to steer your customers toward one of the dishwashers that allow you to secure the unit to the cabinet sides, such as some of the Bosch units. — *Mike Barrett*

Cutting Solid Surfacing

Q. What type of saw blade should I use to remove about an inch from a solid-surface countertop?

A. Many manufacturers, including DeWalt, Everlast, and Forrest, make circular saw blades for cutting solid surfacing. The most efficient blade will have as close to a zero rake as possible. But since every solid-surface cut should be trimmed with a router, saw blade choice is not too important, espccially if you are making only an occasional cut.

Once you've made your cut (about 1/8 inch beyond your final mark), clamp a straightedge to the solid surfacing and rout the edge using a straight bit with a template guide or with a template bit with a bearing above the cutter. — *Tony Pelcher*

Restoring Solid Surfacing

Q. What's the best way to restore the shine on an older solid-surface sink that looks dull?

A. Examine the sink closely to see whether it is scratched or merely dull. A sink that is dull but not scratched can be restored with Soft Scrub (a mildly abrasive kitchen cleaner available in grocery stores) and a Scotch-Brite pad from 3M.

If you can feel any scratches with your fingernail, the first step is to sand out the scratches with wet-and-dry sandpaper, available at any auto parts store. It shouldn't be necessary to use anything coarser than 320-grit. Keep the sink wet while you're sanding. After sanding out the scratches, finish with 400-grit wet-and-dry sandpaper, followed by a Scotch-Brite pad with Soft Scrub. In most cases, restoring a sink should take less than 30 minutes.

In general, dark colors are more likely to show scratches than lighter colors. If the sink is dark, allow it to dry after polishing to see if any scratches remain. If necessary, repeat the process until you achieve the desired results. — *Tony Pelcher*

Suitable Tiles for Outdoor Counter

Q. I'm building an outdoor patio, and the plans call for a ceramic tile counter near the grill. What types of vitreous tile are suitable for outdoor use?

A. The most important factor to consider when choosing a tile for outdoor use is how often the tile will freeze, which depends upon the climate. In southern California or Florida, where an exterior tile installation might freeze only once in its lifetime, almost any tile will do. In a climate that has a frost once or twice a year, a vitreous tile should be fine. If your area gets a frost more than once or twice yearly, however, you need to use an impervious tile. Impervious tiles do not absorb any appreciable moisture that might expand when the tile freezes. Using substandard materials or methods can result in the loss of the tile after only one freeze/thaw cycle.

The tiles should be installed with latex-content thinset mortar and premium-quality latex grout. I recommend installing a sloping subsurface, a drainage layer to allow for runoff, and a crack isolation membrane to absorb differential movement between the tiles and the setting bed.

The correct placement of expansion joints is essential to the performance of the drainage and crack isolation membranes. Without them, an otherwise solid installation will eventually fail. On most tile installations, an expansion joint filled with caulk (instead of grout) is required wherever tiles change direction or meet other materials.

In the case of your counter, you will need an expansion joint of at least 1/8 inch between the grill and the tiles. Since high heat may cause toxic fumes to be emitted from the caulk, this joint slot should not be filled. Your grill's housing may overhang the tiles, hiding the open joint. If not, make certain the tile installer keeps the joint slot open, with tile edges neatly aligned. Resist the urge to butt the tiles right up against the grill. If you leave no room for expansion, when the grill heats up and expands, it could shear the tiles right off the surface.   — *Michael Byrne*

Sagging Ceiling Drywall

Q. When 1/2-inch drywall is installed on ceiling joists spaced 24 inches on-center, will the drywall sag? If the ceiling has blown-in cellulose insulation, will the drywall be more likely to sag than it would under fiberglass batt insulation?

A. I have known 1/2-inch drywall to sag when installed on 2-foot centers. According to the USG's *Gypsum Construction Handbook* (800/874-4968, www.usg.com), to avoid sagging, 1/2-inch drywall installed on a ceiling framed at 24 inches on-center should support no more than 1.3 pounds per square foot of insulation, equivalent to about R-26 for cellulose or R-70 for loose-fill fiberglass. For a better installation, either use 5/8-inch drywall, or first strap the ceiling with 1x3s installed at 16 inches on-center, increasing the allowable insulation load to 2.2 psf. Excessive moisture levels in the ceiling drywall can also lead to sagging.

Ceiling drywall should always be installed perpendicular to the joists, even when they are 16 inches on-center. Drywall is much stronger along the length of the panel because the paper facing provides the strength and the paper grain runs lengthwise.   — *Paul Fisette*

Drywall Orientation in Tall Rooms

Q. I am about to install drywall in a workshop with 9-foot ceilings, and I don't see why I should install it the traditional way, horizontally. Why not buy 10-foot sheets, cut off one foot, and hang the board vertically? This would put all the butt seams where the walls meet the ceiling.

A. I often see drywall hung vertically in garages and workshops, which typically have high ceilings and walls longer than 16 feet (the longest available length of drywall).

For your job, you could also buy 9-foot-long drywall, and then you wouldn't have any waste.

Another option is to use 54-inch-wide drywall, which is made for horizontal attachment on 9-foot-high walls. It's available only in 12-foot lengths, but this may work out fine if you locate the butt seams above doors or above and below windows. The 54-inch drywall is available in both standard 1/2-inch thickness and 5/8-inch fire-code thickness, in case that is an issue.

My first choice would be to use 54-inch board because I prefer taping horizontal seams. Plus, even when you include any butted seams, you may actually end up with fewer lineal feet of seams to tape.

My second choice would be vertically oriented 4x9-foot drywall. If you install it this way, be careful not to place a seam on an outward crowned stud, and double-check your stud spacing to be sure that you won't have to cut off a beveled edge to hit a stud.

My last choice would be to hang 48-inch drywall horizontally. Not only does this approach generate the most lineal feet of seams to tape, but locating the seams is a problem: seams up at the top or low along the floor are inconvenient to tape, while a double seam in the middle is difficult to hide.

For really tall rooms — like the 20-foot living room I'm doing right now — I hang the board horizontally, because otherwise it's nearly impossible to tape. It's standard to put on the stilts or work off scaffolding to tape the joints horizontally around the room, whereas trying to tape a joint that starts 20 feet high at the ceiling and runs down to the floor is nearly impossible.

Another advantage to running the board horizontally in tall spaces is that it allows me to bridge the junction between floors. I'll start with a 2-foot rip at the bottom if necessary so that a full-width board covers the joist area at the floor junction. I also avoid screwing into the band joist, because that's the most likely place for shrinkage and pops.   — *Myron Ferguson*

Preventing Drywall Corner Cracks

Q. I've noticed that whenever I finish drywall, fine cracks start appearing at inside corners a couple of days after the job's done. What is causing the cracks?

A. Assuming that there are no problems with the framing, the most likely cause for the hairline cracks you describe is torn bedding tape. If the finisher allows the corner of his taping knife to cut, or "score," the tape, then a hairline crack will often develop.

Caulking or applying another coat of compound to the crack are temporary fixes at best. The corner should be retaped, and then finished again. Check the corners of your finishing knife to make sure they're not sharp.

Another possible cause is that too much compound is being applied in one pass. Many finishers will apply compound to one side of an inside corner (covering the tape), and allow this first pass to dry before covering the other side of the corner with compound. If the second pass is too heavy, it may shrink back when it dries and leave a hairline crack. Sanding this heavy coat until it just covers the tape will usually eliminate the crack. You should apply only enough compound to inside corners so that after it is lightly sanded, it barely conceals the tape.

It's also important to prefill any gaps in the drywall at corner intersections. If the gaps exceed 3/8 inch, fill them with Durabond (United States Gypsum Co., 800/552-9785; www.usg.com), a low-shrinkage compound that is mixed on site. Durabond is available in grades with setting times from 45 to 90 minutes.

Two other points to keep in mind: If you add water to standard compound, don't exceed the manufacturer's recommendations — generally one pint of water to five gallons of mud. And don't use fiberglass mesh tape at inside corners unless you bed it in Durabond.   — *Mel Hines*

Drywall Nail Pops

Q. We're getting more nail pops in our drywall applications this year than ever before. I've been told that as wood gains moisture, it swells and squeezes the nail out of the nail hole. Are all these nail pops caused by excessive moisture in the house? How can we prevent them?

A. Nail pops are related to changes in the moisture content of your framing, but they're a sign of moisture loss, not moisture gain.

Pops generally result when drywall is fastened to the face of framing material with a relatively high moisture content. As the studs dry, they shrink away from the nail head, leaving the head proud of the drywall surface. The shorter the penetration of the nail into the wood, the less shrinkage along the shaft of the nail, and the less noticeable the nail pop. So one way to minimize the magnitude of pops is to use the shortest nails allowed — 1 1/4-inch nails for 1/2-inch drywall and 1 3/8-inch nails for 5/8-inch drywall.

The best advice, however, is to use dry framing lumber. Do not use lumber that is stamped S-green — that stands for "surfaced-green" and indicates a moisture content greater than 19%. Instead, purchase lumber that is either kiln dried (KD) or S-dry (which indicates a surface moisture content of less than 19%). You must still verify that the lumber has been kept dry in storage and is as close as possible to the in-service moisture content you want (about 12% to 14%). You can verify that with a moisture meter.

Also, it's better to use drywall screws instead of nails, using the shortest screw size permitted.   — *Paul Fisette*

Drywall Callback in Bathroom

Q. We are getting a callback on drywall screws in a bathroom that are showing through the paint and getting darker with age. What might be causing this?

A. It is possible that the heads of the screws are corroding, but this is unlikely if you are using standard corrosion-resistant drywall screws. Stainless-steel screws are best for use in high-moisture areas.

The most probable cause of the problem is condensation on the screw heads. The screws are good conductors of heat and are sunk well into the cold studs on an exterior wall. Since they are insulated only by wood, not by the cavity insulation, the screws can become one of the coldest objects in the room, and a thin film of condensation may form on the walls over the screw heads. This moisture film will attract dirt, forming the dark spots.

To check if dirt is sticking to the wall, wash the wall to see if the spots will come off. While this may seem unusual, I once found a house where the head of every drywall nail was dark, with some other dark streaking. The pilot light on a gas log had been misadjusted and was producing carbon black. The fine carbon particles were attracted to any area that had a condensation film, and the house looked as though there had been a fire. A similar problem often appears in churches, resulting from the greasy carbon produced by candles. — *Henry Spies*

Patching Water-Damaged Plaster

Q. My company has done plaster repair work for many years. After repairing water-damaged walls or ceilings, we occasionally get called back to "fix" an unsuccessful repair. Our second repair attempt involves digging out a very bumpy, chalky substance, and more often than not, we have to repeat this process a number of times until the patch finally takes hold. What causes this reaction on some water damage jobs and not others? What is the most efficient way to deal with this problem?

A. The problem you refer to is caused by efflorescence — salts in the plaster are brought to the surface by the intruding water. The water often causes the magnesium in the lime coat to expand and produce the blisterlike effect you refer to.

Assuming that the water intrusion has been stopped, the first step is to chip away the lime coat at the affected area. There is a good chance that the bond between the finish lime coat and the plaster base coat has been weakened.

When faced with this situation, I use a wire brush to scrub away any loose base-coat particles, then apply a coat of Kilz (Masterchem Industries, LLC, 866/774-6371; www.kilz.com) to the base-coat plaster. Kilz is an alkyd-based sealer, primer, and stain blocker. The plaster base coat must be completely dry before application. Next, I apply a coat of Durabond (United States Gypsum Co., 800/552-9785; www.usg.com). Durabond is a fast-setting, low-shrinkage compound with tenacious bonding qualities and accelerated setting times. Finally, I apply a skim coat of ready-mixed all-purpose joint compound. After a light sanding, the repair is ready for painting. *— Mel Hines*

Metal Lath Ceiling Demo

Q. I need to remove a metal lath and plaster ceiling in a finished basement. The existing partitions were installed after the ceiling, and I want to disturb the finished walls as little as possible. What tools and techniques should I use for this demolition job?

A. In the center of the ceiling, where there is no particular need to be fussy, you can beat out the plaster fairly quickly. When you get to the edges, use a large screwdriver or chisel to chip away the plaster more carefully, so as not to damage the adjacent surfaces. You want to leave a clean edge in the corners, if possible. Where necessary, use metal snips to cut the lath. *— Ron Webber*

Clear Coats for Interior Woodwork

Q. What is the best clear wood finish to use on interior doors and cabinets that are exposed to strong sunlight?

A. For previously unfinished wood, a urethane finish would be the best option. Urethanes hold up well to ultraviolet light. However, urethanes tend to be brittle and can crack if the surface is subject to impact.

For wood that has already been coated with a clear finish, your best bet is an alkyd-based varnish, especially if you're not sure what the original finish is. Alkyds are compatible with a wide range of finishes, whereas the solvents in urethane may cause some clear finishes to bubble. A very light coat of the alkyd-based varnish might yellow slightly, but this is not noticeable on most surfaces. Any new finish should be tried in an inconspicuous spot to be sure it will dry properly and won't produce any adverse effects. — *Henry Spies*

UV Blockers for Clear Finishes

Q. Most species of trim that we install become darker as they age. I've read that this darkening is caused by exposure to the ultraviolet (UV) rays in sunlight. Occasionally, a customer will complain about this darkening. Is there a clear finish available that will block this UV action?

A. Exposure to sunlight will cause most species of wood to darken, but the darkening is not caused exclusively by UV radiation. Color changes occur when the pigments within the wood oxidize. This oxidation is caused not only by the "free radicals" generated by UV radiation, but also by the presence of oxygen. The UV absorbers found in many finishes inhibit or partially block UV radiation from reaching the wood, but since no finish is totally impermeable to oxygen, the color change will still occur, though more gradually.

To postpone the inevitable, I recommend using a film-forming finish, which blocks oxygen penetration much more effectively than a penetrating finish. Make sure the finish contains a UV "absorber," not a UV

"inhibitor." UV inhibitors are used to protect the finish film from UV degradation, and they do not protect the wood pigment from UV exposure as effectively as UV absorbers.
— *Tom Brown*

Finishing Water-Stained Wood

Q. I have an unfinished mahogany door that has water stains. How can I get rid of these stains before finishing?

A. Unfortunately, there's no way to bleach out that stain. You'll either have to sand down to new wood if possible, or try to match or mask the tone of the stained area with another stain or dye over the entire door.
— *Robert Sanders*

Painting Over Smoke-Damaged Walls

Q. Is there a paint or sealer that will encapsulate the smell of smoke on fire-damaged walls?

A. Several paint manufacturers have special products for sealing in smoke. One is Lok-Tite Stain & Tannin Blocker (M.A.B. Paints, 800/MAB-1899; www.mabpaints.com).
— *Henry Spies*

Painting Popcorn Ceilings

Q. Several of my customers have "popcorn" ceilings that are old and dirty. Is there any way of painting them short of spray-painting?

A. I always spray-paint popcorn — or blown-on acoustic — ceilings. When you roll them, the material breaks loose and clogs the roller. If you have to roll them, look for a special textured foam roller cover designed for acoustic ceilings. These covers have slits and cross-hatchings that allow the foam to better conform to the irregular ceiling surface. The idea is to apply the paint with the least possible pressure to prevent the ceiling material from breaking loose.

Cutting in the corners can also be tricky. If you try to cut in with a brush, you will have to work hard to avoid lap marks. It may be faster to mask the walls and roll right up to the edge.

It will take several coats of paint on the entire ceiling to produce an even finish. Be advised that the water in the paint will wet the popcorn surface, causing it to roll off with the roller. Apply the first coat and allow it to dry thoroughly before you try to backroll or apply additional coats.

One reason people often want to repaint acoustic ceiling is to hide water stains caused by plumbing or roof leaks. But those stains are water-soluble and will telegraph through new paint. To prevent that, always use several coats of a shellac primer to lock in the stain and prevent it from bleeding through.

While you're at the paint store, inquire about ceiling paints that are engineered especially for acoustic ceilings. Such paints have less resin density than standard smooth-wall paints and will help maintain the acoustic qualities of the rough, textured surface.   — *Dan Greenough*

Black Stripes on a Cathedral Ceiling

Q. I have been asked to determine the cause of black shadow marks on a cathedral ceiling in a new home. The black marks correspond to the bottom sections of the rafters, near where the rafters meet the wall. The black shadows do not resemble stains from water leaks. There is a gas fireplace on one of the outside walls; the fireplace has been tested for leakage, and it checked out fine. The owner burns a large candle, which sits on the fireplace mantle. Could the ceiling shadows be caused by soot from the candle?

A. Reports are increasing of problems with black stain deposition, sometimes called "ghosting" or "sooting," on interior surfaces. Since tiny particles called "particulates" are attracted to cooler surfaces, it is not uncommon to see black staining following the framing members behind drywall. Common sources range from furnaces and water heaters to fireplaces, candles, outdoor cooking grills, and even automobile tires.

Although you say the fireplace has been tested for leakage, it's important to verify that the fireplace was tested under pressure. Some units appear to be sealed, but can leak significantly under negative pressure. Is there a standing pilot light? Does it impinge on any of the logs? Has the gas pressure been tested? Does the orifice match the fuel being burned? These items should all be checked, since we have found that any one of them might be the source of particulate matter.

We also know that many types of jar candles, imported candles, highly fragranced candles, and cans with improper wicks can burn dirtier than other types and can be the source of black stain deposition. Stains can also result from improper burning practices, including failing to trim the wick, burning a candle near a draft (for example, near a heating register or ceiling fan), and burning a candle for too many hours. If you suspect that a candle is the cause of the staining, tell the homeowner to stop burning the candle for a few months. If the stains return to the ceiling after repainting, you know the source is from something else.

Covering the stains requires a good-quality sealer, such as Kilz, prior to painting. Otherwise, the stain will bleed back through the paint.

Because particulates are attracted to cooler surfaces, there is a strong likelihood that the ceiling insulation is poorly installed — perhaps because of voids or gaps along the edges of the insulation. The owner of the house may wish to improve the ceiling insulation.

Ghosting, or sooting, on carpeting and furnishings is also common in today's tighter houses — especially tight homes with insufficient ventilation — where particulates have more opportunity to deposit, instead of being flushed away by regular air changes.

The driving force for stains along the edges of carpets is typically pressure caused by mechanical fans and/or the stack effect. Where escaping air goes up the wall cavity to the attic, the carpet serves as a filter, getting stained as it scrubs the air of contaminants. On the other hand, negative pressure in the house could be causing attic air to filter down the walls, with the carpet again serving as a filter.

Pressure mapping of the house by a qualified technician can pinpoint the pressure dynamics the house experiences while fans are operating. From there, you can begin to trace likely sources for the stain.

— Frank Vigil

Finishing Crown Molding

Q. I'm installing a lot of crown molding in a finished home, and wondering whether it's better to prepaint the crown and just touch up the nail holes or paint it once it's in place?

A. Prepainting might seem like a good idea at first, but you'll get a better job in the same or less time if you paint the crown in place. The fact is, the walls and ceiling always have some irregularities and the crown may be warped or cupped. Inevitably there will be gaps where the crown meets the walls and ceiling, and for a professional-looking job, these gaps must be caulked. The caulking, plus any spackling at nail holes and joints, must be painted. You might try cutting in just the caulk lines and touching up the spackling, but since neither latex nor alkyd enamel patches well, the areas you painted in place would likely show up.

I would prime and sand the crown, then install it. Spackle, then sand smooth all nail holes, corners, and joints. Spot-prime all spackling. Fill any gaps behind or above the crown that are larger than $1/16$ inch. Caulk the entire length of the crown, at both ceiling and walls. Top-coat with two coats of enamel, then cut into the crown with the appropriate ceiling or wall paint. Make sure your customer understands that all newly applied paints will not exactly match the color of the existing paints and may visibly affect the texture of the existing walls.　　— *Robert Sanders*

Sealing Cedar in Closets

Q. Does T&G cedar paneling for closets need to be sealed? If so, what should I use?

A. Cedar paneling in closets should never be sealed. The purpose of the cedar paneling is to release the aroma and volatile compounds in the cedar into the closet to repel moths. As the volatiles decrease with age, the effects of the cedar can be restored with a light sanding to expose new wood, or by applying a coating of cedar oil.　　— *Henry Spies*

White vs. Yellow Glue for Interior Trim

Q. For gluing joints in interior wood trim that will never get wet, is white carpenter's glue as strong as yellow carpenter's glue?

A. Yes, and both glues will form a bond stronger than the wood itself. The main difference between the two glues is in application characteristics. Both glues belong to a class of glues called PVAs, because they are based on a resin called polyvinyl acetate. White glues like Elmer's were the first glues of this type to be introduced. While they are perfectly fine for gluing wood (and a host of other porous materials), they do not sand well, tend to run, and have low initial tack. Yellow glues were developed to address these problems. Yellow glues have a higher solids content (they are thicker than white glues and sand better) and have tackifiers added to speed assembly time. To distinguish these improved glues from the older white glues, manufacturers added a yellow color and a meaningless designation, calling them "aliphatic" resin glues. If you think of yellow glue as white glue with a higher solids content, some additives, and yellow color, you're on the right track. Yellow glues are definitely easier to use and have better application characteristics but are no different in strength from white glues. — *Jeff Jewitt*

Disassembling Yellow Glue Joints

Q. After a wood joint glued with yellow carpenter's glue has cured, is there any way to disassemble the joint?

A. Both yellow and white PVAs will soften and lose some of their grip if you saturate the joint with hot vinegar. Use a syringe to introduce the hot vinegar. If necessary, drill 3/32-inch holes in an inconspicuous spot to get the hot vinegar into the joint. Let it sit for at least 30 minutes, and rewet the joint if it looks dry. Then try to wiggle the joint apart. Do not whack it with a hammer (at least, not very hard), as you run the risk of breaking the wood. — *Jeff Jewitt*

Best Glue Without Clamping

Q. What's the best glue to use for a wood joint when it is impossible to clamp the joint?

A. If it's a structural joint, the best glue is epoxy, because it has the best gap-filling ability, without compromising overall joint strength. You still have to bring the two surfaces as close together as you can, so I'd use five-minute epoxy and hold the parts together. For nonstructural gluing (like gluing a rosette), my first choice is always cyanoacrylate glue, used with an accelerator. Use a thick-viscosity cyanoacrylate, like Quick-Set (available from Rockler Woodworking; 800/279-4441, www.rockler.com) and apply it to one surface and the activator to the other. The big downside of the cyanoacrylates is that you need to get the fit right immediately, as the bond is instant and fussing with the fit will compromise the strength.

A clamped joint will always be stronger than one held together with a dab of hot-melt. For the best possible bond, the pieces being glued need to be held in intimate contact while under pressure. Any other method will result in a bond of less strength. This is due to the thickness of the resulting bond. As a rule, a thick glue line is a weak glue line.

— Jeff Jewitt and Jeff Pitcher

Best Application for Hot-Melt Glue

Q. Some builders use hot-melt glue for temporary jigs. Can hot-melt glue also be used for a permanent bond? What types of applications make sense for hot-melt glue?

A. The use of hot-melt as a "permanent" adhesive has been debated. Personally, I don't use it for permanent bonds, but the furniture industry does, using it for a whole array of purposes, including edging and face-lamination of vinyl. Hot-melt doesn't penetrate and "wet" a wooden surface the way other glues do, so the bond it forms isn't as rigid. Applications for which I would consider using hot-melt include gluing fabric and applying melamine edging to particleboard. One thing is certain: the consumer grades of hot melt glue are far different from the industrial grades used by the furniture industry. *— Jeff Jewitt*

Will Freezing Ruin Yellow Glue?

Q. If a joint assembled with yellow carpenter's glue freezes before it cures, will the joint lose its strength? What is the best exterior wood glue to use in below-freezing temperatures?

A. To answer your first question: absolutely. The minimum application temperature for PVA glues, including yellow glue, is 50°F. While the unused glue can endure several freeze/thaw cycles, a joint with frozen glue will not develop a bond that's as strong as that in warmer temperatures. The glue will "chalk" at temperatures between 30°F and 40°F.

Try it yourself — glue some wood together and stick it in a freezer. The next day, you should be able to pull the joint apart by hand.

Polyurethane glues, like Excel or Gorilla Glue, are the best exterior wood glues to use in colder temperatures. Most water-based products, like cross-linking PVAs, have a minimum-use temperature (some as high as 60°F). Because reactive polyurethanes are 100% solids, there is no concern with minimum-use temperatures. In fact, tests have shown that the bond strength with some polyurethane adhesives actually increases as the temperature drops. Although most polyurethane glue manufacturers recommend that their glues be applied above a minimum temperature (usually 32°F or 40°F), they can probably be used at lower temperatures.

— *Jeff Jewitt and Jeff Pitcher*

Gluing to End-Grain

Q. What's the best wood glue for end-grain applications — miter joints, for example?

A. This is one of the most difficult joints to glue. Epoxies and reactive hot-melts are better than water-based wood glues. (Reactive hot-melt glues such as Titebond HiPURformer differ from typical hardware-store hot-melt glues.) Since the end grain of wood is very porous, water-based wood glues are easily absorbed into the grain, leaving little glue for bonding. A method for combating this is to thin some of the glue with water and use that to seal the end grain, then glue up as normal after an hour or so. There are some wood sizing (sealing) products on the market that work to seal the end grain, such as Glue Size (Custom-Pak Adhesives, 800/454-4583; www.custompak.com).

— *Mark A. Roberts*

Gluing Miter Joints

Q. In my area of South Carolina, changes in humidity often cause miter joints to open up over time. I plan to use biscuits to keep my miters tight. Would I be better off using epoxy instead of yellow glue when gluing up the biscuits? Should I check all my lumber with a moisture meter?

A. To many carpenters, "moving" miter joints can be a vexing problem. The customer complains about unsightly gaps in trim joints that fit perfectly when they were first installed. While it's true that open miters (and other defects) can be caused by changes in humidity, other factors also affect joint movement.

As a general rule, the moisture content of trim should be about 8%. This figure increases to 11% in damp coastal areas and drops to 6% in arid climates. Variations in desired moisture content of 1% to 2% are acceptable. You should definitely use a moisture meter. I use a no-frills version with a range of 6% to 30%, made by Delmhorst Instrument Co. (877/335-6467, www.delmhorst.com).

Wood movement is proportionate to board width. Trim wider than 4 inches is more likely to cause problems. Whenever possible, build up wide casings from narrower profiles.

Some wood species are more prone to movement (beech and maple, for example), while woods like cherry, red oak, and white pine are more stable. Quartersawn lumber of any species is more stable than flat-sawn, but is often expensive and difficult to find.

Biscuits are an excellent way to ensure that a miter joint "stays put," but they should not be used to compensate for improperly dried lumber. When gluing biscuits, epoxy offers no advantage over yellow (aliphatic resin) glue. The moisture in the yellow glue causes the compressed biscuit to swell, closing the gap between the slot and biscuit, and strengthens the joint.
— *Michael Poster*

Polyurethane Glue With Wet Wood

Q. Is it possible for wood to be too wet for polyurethane glue?

A. You can expect polyurethane glue to cure properly in woods with a moisture content of up to 25%. Moisture has to be present for polyurethane glue to cure, so the moisture content of the wood to be glued should be at least 8%. While there is no upper limit per se for the glue to cure, I'd avoid gluing wood with a moisture content higher than 25%, regardless of the glue. Wood with such a high moisture content will shrink appreciably, and the chances of joint failure are high.

— Jeff Jewitt

Polyurethane Glue Cleanup

Q. What is the best way to clean polyurethane adhesive from skin? How about from carpeting?

A. The best idea is to wear gloves and not get it on your skin to begin with. I use cheap vinyl gloves from Grainger's. If polyurethane glue does get on your skin, wipe it immediately with a waterless hand cleaner. After it dries, the only way to remove it is to use an aggressive pumice-based hand cleaner, which will remove most of the glue (and part of your skin). The dried glue will wear off in about a week. On carpeting, the last thing I would do is try to clean it by wiping it with anything containing water. Naphtha should remove the bulk of it, without initiating curing. *— Jeff Jewitt*

Wood Glue: Okay After Freezing?

Q. We use a yellow carpenter's glue for wood-to-wood gluing chores. Every winter, I invariably leave the container in my truck overnight, and the glue freezes. Can this glue be thawed and used after it has frozen?

A. "Carpenter's" glue generally comes in two flavors: polyvinyl acetate (better known as white glue) and aliphatic resin (yellow glue).

When I contacted Franklin International, manufacturer of Titebond wood glue, they said that their white and yellow glues can undergo five freeze/thaw cycles before they should be discarded. Frozen glue often thaws to a thicker consistency, and Franklin mentioned that up to 5% water can be added (by volume or weight) to thin their glues.

Custom-Pak Adhesives notes that not all glues are freeze/thaw stable, but those that are should be allowed to thaw completely and stirred thoroughly before use.

If you're uncertain about the number of freeze/thaw cycles a glue has undergone, the Department of Forestry at the University of Wisconsin-Madison offers this advice: if the glue appears to be the same in color and consistency after thawing, chances are it can be used.

But keep things in perspective. By using questionable glue, are you risking costly callbacks in an effort to save a few dollars? If the glue is lumpy or differs in consistency, then toss it. *— Carl Hagstrom*

Anita Aikey-Socinsky is a master fabricator and owner of the custom natural stone business, A&M Stoneworks, Inc., in Colchester, Vt.

Sal Alfano, a former builder, is the editorial director of *JLC, Coastal Contractor, Remodeling, Replacement Contractor*, and *Upscale Remodeling* magazines.

Richard Allen, Jr., is a staff engineer at the Brick Industry Association.

Brent Anderson, P.E., is president of BA Associates, specialty contractors and consulting engineers in Fridley, Minn.

Steve Andrews is the author of *Foam Panels and Building Systems*.

Julius Ballanco, P.E., is president of J.B. Engineering and Code Consulting in Munster, Ind.

David Ballantyne is a project engineer at Icynene, Inc.

Mike Barrett is the owner of Vermont Precision Stone in South Burlington, Vt.

David Beal is a research analyst at the Florida Solar Energy Center.

James Benney is a member of PaintCraft Associates, a guild of finishing experts in the San Francisco Bay area.

Eric Borden is owner of ESB Contracting in Forked River, N.J.

Robert Bouchet is an engineer with the Simpson Strong-Tie Co. in Pleasanton, Calif.

David Bowyer is a designer and sales manager for Barlow Custom Builders, an Edwardsburg, Mich., remodeling firm.

Terry Brennan is a building researcher in Oriskany, N.Y.

Tom Brewer is a roofer in Hallstead, Pa.

Howard Brickman is a flooring contractor and consultant in Norwell, Mass.

Tom Brown is a former wood finishing consultant in Ft. Myers, Fla.

Joseph Bublick is a roofer in Toledo, Ohio.

Stephen Bushway is a mason and venting system specialist in Plainfield, Mass.

Michael Byrne is an expert tilesetter and consultant in Los Olivos, Calif., as well as a contributing editor to *JLC* and author of many *JLC* articles and the book *Setting Tile*.

Steve Campolo is vice president for engineering at Leviton.

Mike Casey is a licensed plumbing contractor in California and Connecticut and coauthor of *Code Check* (Plumbing and HVAC).

Rex Cauldwell is a master electrician in Roanoke, Va., a frequent contributor to *JLC*, and the author of several books, including *Wiring a House*.

Bill Clinton is a hydronic heating contractor in Sonoma, Calif.

Rob Corbo is a general contractor in Elizabeth, N.J.

Dave Cunningham is a foundation repair contractor in Independence, Mo.

John Curran owns RSI General Contractors, a roofing, siding and insulation company in Syracuse, N.Y.

Ted Cushman is a former associate editor at *JLC*. He writes frequently on construction-related topics from Great Barrington, Mass.

Patricia Davidson is with the Floor Covering Installation Contractors Association.

Bruce Davis is a senior consultant at Advanced Energy in Raleigh, N.C.

Christopher F. DeBlois, P.E., is a structural engineer with Palmer Engineering Co. in Tucker, Ga.

Clayton DeKorne is a former builder and was founding editor of *Tools of the Trade* magazine. As founding editor of *Coastal Contractor* magazine, he writes and edits on a wide variety of construction-related topics from Brooklyn, N.Y., and Burlington, Vt.

Henri deMarne of Waitsfield, Vt., consults on building technology, and as a former contributing editor has written many articles for *JLC*.

Dan Dolan, P.E., is associate professor of Wood Engineering at Virginia Tech.

Jeri Donadee, vice president of H.B. McClure Co., is a heating and cooling contractor in Harrisburg, Pa.

Brad Douglas is director of engineering at the American Forest & Paper Association.

Larry Drake, of Hyram, Utah, is executive director of the Radiant Panel Association.

Curtis Eck is a Boise-based technical representative for Trus Joist MacMillan.

John Edgar is senior technical services manager at Sto Corp., a manufacturer of EIFS systems.

Bill Eich owns Bill Eich Construction in Spirit Lake, Iowa.

Bill Feist, a consultant and teacher on wood weathering and exterior wood finishing, was a research chemist at the Forest Products Lab in Madison, Wis., for 30 years.

Myron Ferguson is a drywall contractor in Galway, N.Y., and the author of *Drywall: Professional Techniques for Great Results.*

Ed Fillbach is a former painting contractor in Bozeman, Mont.

Paul Fisette is director of the Building Materials and Wood Technology program at the University of Massachusetts in Amherst and a contributing editor to *JLC*.

Tim Fisher was formerly the field engineering editor for Aberdeen's *Concrete Construction* magazine.

Don Fugler is senior researcher at Canada Mortgage and Housing Corp.

Rocky Geans, president of L.L. Geans Construction Co. in South Bend, Ind., is a member of the board of directors for the American Society of Concrete Contractors

Gary Gerber is owner of Sun Light and Power, in Berkeley, Calif.

Dave Gobis is executive director of the Ceramic Tile Education Foundation.

Chuck Green, a NARI Certified Remodeler and Certified Lead Carpenter, owns Four Corners Remodeling in Ashland, Mass.

Dan Greenough is a painting and finishing contractor in the San Francisco Bay area.

Clayford Grimm is a consulting architectural engineer from Austin, Texas.

Mike Guertin is a builder and remodeler in East Greenwich, R.I., and a member of the JLC Live! construction demonstration team.

Rob Haddock is with the Metal Roof Advisory Group, Ltd.

Carl Hagstrom, a mason and builder for 20 years, owns Hagstrom Contracting in Montrose, Pa. and is a contributing editor to *JLC*.

Ron Hamilton is owner of Hamilton General Contracting in Saylorsburg, Pa.

Jim Hamrick is an architect in York, Pa., and an instructor at Stevens State School of Technology in Lancaster, Pa.

Jennifer Hause is an engineering scientist with the National Small Flows

Clearinghouse at the National Environmental Services Center of West Virginia University.

Mel Hines owns Atlanta Pro-Serve, a ceiling and wall repair service in Atlanta, Ga.

Duffy Hoffman is owner of Hoffman Painting and Refinishing, Inc., in Pipersville, Pa.

Martin Holladay, a former associate editor at *JLC*, is the current editor of *Energy Design Update.*

Will Holladay, a longtime framer, is the author of *A Roof Cutter's Secrets to Framing the Custom House* (JLC Books).

Bonnie Holmes is director of education at the National Wood Flooring Association.

Pat Huelman is an associate professor at the University of Minnesota.

Don Jackson is a former builder and chief editor at *JLC.*

Daniel C. Jandzio is branch manager at Coastal Metal Service, a manufacturer and supplier of commercial and residential roofing and metal products.

Jeff Jewitt runs Homestead Finishing Products, a restoration and finishing supply company. He is the author of *Great Wood Finishes*, and has made two full-length feature videos, "Coloring Wood" and "Applying Top Coats."

Stephen Jordan is a Rochester, N.Y., architectural conservator, preservation specialist, and author.

Charlie Jourdain is a finishes specialist at the California Redwood Association.

Tony Jucewicz is a stone mason in Riegelsville, Pa.

Redwood Kardon is a building inspector for the City of Oakland, Calif., and author of *Code Check: A Field Guide to Building a Safe Home.*

Rick Karg is an energy management consultant in Topsham, Maine.

Mark J. Katuzney is owner of Mar-Kay Siding and Roofing Co. in Yalesville, Conn.

David Keener is an engineer for LSA, Inc., in Lancaster, Pa.

Ed Keith is a senior engineer at APA — The Engineered Wood Association.

Sean Kenney owns and operates Sean M. Kenney Electrical in Amesbury, Mass.

Mike Keogh is a specialist in energy conservation and roof ventilation from Campbellford, Ont.

Al King formerly operated a heating and plumbing contracting business in Perkasie, Pa.

Mark Knaebe is a Forest Products Technologist at the USDA Forest Products Laboratory.

Bob Kovacs is moderator of the Estimating and Markup forum at jlconline.com.

Moncef Krarti is a professor of architectural engineering at the University of Colorado in Boulder.

Mike Lacher was formerly a technical services manager in the Insulation Group at CertainTeed Corp.

Art Laurenson is a Providence, R.I., firefighter and former assistant deputy state fire marshal.

John Leeke is a preservation consultant from Portland, Maine.

Eric Lewis is the former owner of Spectrum Electrical Services in Montrose, Pa.

Terry Love is a plumbing contractor in Bellevue, Wash.

Joe Lstiburek is a principal of Building Science Corp. in Westford, Mass., and author of the *Builder's Field Guide.*

Don Marsh was formerly a project manager with Dufresne-Henry Consulting Engineers in Montpelier, Vt., and the media services representative for the Portland Cement Association.

Felix Marti has been a carpenter, designer, and general contractor for over 40 years in the Ridgway, Colo., area.

Harrison McCampbell is an architect and roofing consultant in Nashville, Tenn., specializing in construction defects.

Patricia McDaniel is owner of Boardwalk Builders in Rehoboth Beach, Del.

Scott McVicker is a structural engineer in Half Moon Bay, Calif.

Tom Meehan is owner of Cape Cod Tile Works in Harwich, Mass.

Jay Meunier ran his own concrete contracting business for many years, and is now an estimator with Pizzagalli Construction in Burlington, Vt.

Cyrus Miller is a production supervisor for Common Vision Inc. in Hamden, Conn.

Doug Mossbrook is president of Eagle Mountain HVAC in Canandaigua, N.Y.

Bruce Nelson is a senior engineer for the Minnesota Department of Public Service.

Joel (Ned) Nisson is the former editor of *Energy Design Update* and a former consultant on energy-related construction. He now resides in Northampton, Mass.

Martin Obando is a master roofer and director of applications specifications for the Cedar Shake and Shingle Bureau.

Chip O'Rear works in the tile industry and was the assistant executive director of the National Tile Contractors Association.

Bill Palmer, P.E., is editor of *Concrete Construction* magazine.

Danny Parker is a research scientist at the Florida Solar Energy Center.

Tony Pelcher is a solid-surface fabricator at the Top Shop in South Burlington, Vt.

Kevin Pelletier is a sales representative for Wood Structures Inc., a Boise Cascade distributor in Biddeford, Maine.

Richard Piper is an architect and EIFS specialist from Brighton, Mass.

Jeff Pitcher is vice president of CP Industries in Clifton, N.J., an affiliate of CP Adhesives in Newark, Ohio.

Michael Poster is president and co-owner of WOODWEB.COM

Michael Purser is a second-generation floor finisher and owner of Rosebud Floors in Atlanta, Ga.

Stephen Quarles is cooperative extension advisor on wood durability at the University of California in Richmond, Calif.

Robert Randall, P.E., owns Randall Millenium Homes, LLC, in Mohegan Lake, N.Y., whose mission is to market and sell zero-energy homes.

Jim Reicherts is with the United States Gypsum Company.

Kelly Reynolds is principal of Kelly P. Reynolds & Associates, a firm specializing in code interpretation and plan review, with offices in Chicago and Phoenix.

Don Richardson is president of the U.S. Division of Romaro Structures, a truss manufacturer.

Bruce Richgels is an account representative for Waterproofing Inc., a Minneapolis-based company specializing in foundation waterproofing.

Mark A. Roberts is a senior technical specialist in Franklin International's Construction Products Division.

Bill Rose is a research architect at the University of Illinois at Urbana-Champaign.

Marc Rosenbaum, P.E., of Energysmiths in Meriden, N.H., designs and engineers high performance buildings.

Robert Sanders is a remodeling contractor in Pasadena, Calif.

Craig Savage, a former builder, editor at *JLC*, and director of the JLCLIVE training shows, is the author and co-author of several books on construction and the current chief marketing officer for Building Media Inc.

George Schambach is a former siding contractor in Deposit, N.Y.

Fred Seifert, Sr., passed down the family business to his sons, and now works for them at Seifert Brothers Construction in Mattituck, N.Y.

Mike Shannahan, master carpenter, owns and operates a restoration contracting business in La Porte, Texas.

Andy Shapiro is an energy and sustainable design consultant in Montpelier, Vt.

Robert Shuldes, P.E., is an engineer with the Portland Cement Association.

John Siegenthaler, P.E., owns Appropriate Designs, a building systems engineering firm in Holland Patent, N.Y.

Chuck Silver designed energy-efficient homes for many years and conducted training seminars for builders in New Paltz, N.Y.

Bill Rock Smith, a former contractor, is a building consultant in Latham, N.Y.

Ken Smith owns and operates Smitty's Flooring Specialties in Randolph, Vt.

Scott Smith is vice president of the nail manufacturer Prime Source Building Products.

Dr. Stephen Smulski is a wood scientist and president of Wood Science Specialists in Shutesbury, Mass.

Mark Snyder is a builder and consultant who divides his time between Massachusetts and Vermont.

Henry Spies, formerly with the Small Homes Council-Building Research Council of the University of Illinois, is a building consultant in Champaign, Ill.

Cliff Thomas is the owner of Hands-On Electric in Santa Fe, N.M.

Steve Thomas has worked in residential construction since 1987, specializing in stucco, EIFS, brick, cultured stone, and block. He currently is an estimator for a masonry firm in Dublin, Ohio.

Jon Tobey is a painting contractor in Monroe, Wash.

Jeff Tooley is owner of the Healthy Building Company, based in Bear Creek, N.C.

David Utterback is the owner of TimberTek Consulting in Overland Park, Kansas, and is a building inspector for the city of Lenexa, Kansas.

Frank Vigil is a former building science specialist in Las Vegas, Nev.

Ron Webber is owner of ProCoat Systems in Orange, Calif.

Martin Weiland is manager of technical services at ASHRAE.

Bob Werner is an applications sales specialist at CertainTeed Corp.

Phil Westover, P.E., is a former consulting engineer in Winchester, Mass.

Frank Woeste, P.E., is a professor emeritus at Virginia Tech University in Blacksburg and a frequent contributor to *JLC*.

Bruce A. Wooster owns and operates William R. Wooster & Sons in Crisfield, Md.

Dave Yates owns and operates F. W. Behler, Inc., a plumbing, heating, air conditioning, and solar contracting firm established in 1900 and located in York, Pa.

Bill Zoeller is a senior architect with Steven Winter Associates, a building systems consulting and research firm based in Norwalk, Conn.

Index